Water Conservation in Urban Households

Water Conservation in Urban Households

Role of Prices, Policies and Technologies

Sonia Ferdous Hoque

IWA PUBLISHING

Published by **IWA Publishing**
Alliance House
12 Caxton Street
London SW1H 0QS, UK
Telephone: +44 (0)20 7654 5500
Fax: +44 (0)20 7654 5555
Email: publications@iwap.co.uk
Web: www.iwapublishing.com

First published 2014

British Library Cataloguing in Publication Data
A CIP catalogue record for this book is available from the British Library

ISBN 9781780405476 (Paperback)
ISBN 9781780405483 (eBook)

Contents

Chapter 5

References

List of figures

List of tables

List of boxes

About the author

Sonia Ferdous Hoque is currently a postgraduate research student (PhD) at the School of Earth and Environment at the University of Leeds, UK. She has completed her MSc in Environmental Management from the National University of Singapore (NUS) and BSc in Environmental Science from North South University, Bangladesh. She has received the Shell Gold Medal and Prize (2012) and the Chancellor's Gold Medal (2010) for being the most outstanding student of her masters and bachelors cohort respectively. Prior to joining her PhD program, Sonia worked as a Research Associate at the Institute of Water Policy, NUS (2011–2013), where her research focused on roles of private and public sector in urban water governance and water and wastewater tariffs in urban domestic and non-domestic sectors. She also served as a Research Officer at the Bangladesh Institute of Development Studies (2009–2010), where she was involved in research on climate change adaptation in Bangladesh. Her e-mail address for correspondence is soniahoque@gmail.com

Preface

Domestic water consumption is not just a specialised field of research for economists, sociologists, psychologists, engineers or experts from other disciplines. It is an issue that is closely related to the daily lives of the entire human population – we all use water everyday for various purposes, such as drinking, bathing, laundry, dish washing, sanitation and gardening. While some of us are fortunate to have an uninterrupted supply of clean piped water within our household, many others have to struggle everyday in search of water to meet their very basic needs. Establishing water as a human right in international conventions and national constitutions and policies often do not translate into action in the real world. The problem is not related to the fact that there is not enough freshwater in the world to meet the demands of the growing population. The challenge arises from unequal distribution of water between regions and countries and the improper management of the available local resources. In many cities, under-pricing of water, aging infrastructure and poor institutional and financial capacity, coupled with burgeoning population and expansion of city boundaries, leave a large proportion of the urban population deprived from adequate water supply. In others, despite the presence of good governance, technological and financial resources, water security remains a challenging issue due to lack of sustainable local water sources. Political tensions arise when water has to be transferred from one administrative or catchment region to another, and water conflicts break out when private companies fail to meet their contract obligations and satisfy their consumers and regulators. In addition, climatic variabilities are increasing the uncertainties of water sources.

All these challenges highlight the importance of proper management of the available water sources and the water supply systems. Conservation of water in urban households is one of the key measures to ensure long-term water security and sustainability of water resources. Unlike common belief, conservation does not necessarily mean sacrificing satisfaction or performance in resource use; it means achieving the desired task using less amount of the resource in question. Water conservation entails changes in attitudes and behaviours in everyday water consumption activities. While this has personal benefits in terms of monetary savings, it also has wider economic and environmental benefits. Pro-environmental behaviours is much more common in case of other resources such as energy consumption for example. There is now wider awareness among the general population to adopt simple behaviours like switching off lights before leaving the room. Energy generation is more significantly linked to greenhouse gas emissions and the switch to greener alternative sources is receiving much focus. However, despite being a more vital resource than energy, issues of water conservation often do not receive much attention. Is it because water is much cheaper than energy and do not have noticeable impacts on our pockets? Is it because there is less awareness among the general population?

This book is aimed to address the issues of water conservation from a holistic perspective and includes the standpoints of governments, water utility managers, researchers and most importantly, the views of the general population. It is an attempt to consolidate secondary information from academic and grey literature, that are currently available in scattered forms and locations. With reference to case studies of different cities and the findings of several published research studies, this book will act as a one-stop repository of information on all aspects of water conservation in the urban residential sector. The limited use of technical jargons or mathematical elements will make it an easy read for policy makers and the general population who may not be specialized in this area. It can also serve as a starting point for students and researchers who aim to work in this field.

Chapter 1

Water conservation: The answer to water shortage in the 21st century

1.1 THE CHANGING WATER PARADIGM

Access to freshwater supply has always been and will be an indispensable component of human survival and economic development. During the twentieth century, exponential population growth, increased urbanisation and rapid economic development served as the three most important drivers of developing and expanding supplies of freshwater. Between 1900 and 2000, the world's population grew from 1.6 billion to over 6 billion and fresh water withdrawals increased from an estimated 580 km^3/year to about 3700 km^3/year (Gleick, 2000a). Water resource planners and engineers exclusively focused on investing in large scale infrastructure such as dams, reservoirs, treatment plants and distribution pipelines to tap water from nearby surface and groundwater sources. Water resources development in the past century completely ignored the hidden costs of environmental loss, such as the destruction of ecosystems, dislocation of human population, inundation of cultural sites and disruption of sedimentation processes. Economic analyses were generally done with incomplete information. Water prices have been extremely low compared to the cost of extraction, treatment and distribution, and most capital expenditures have been provided by the governments and donor agencies. According to conventional wisdom, the pressing water problem was to close the gap between existing and projected water supply and demand. In most cases, future water consumption estimates were assumed to be direct functions of population size and economic wealth, and as these factors would grow, so would water consumption. However, over time these projections have often proved to be wrong and cities have managed to reduce their per capita water usage and even total water consumption through careful planning and wise allocation of resources (see Box 4.1).

Beginning in the late 1960s and early 1970s, the ecological and political costs of building large-scale water infrastructure became more apparent and the environmental movement began to challenge proposals for new dams. The economic costs of the traditional water path have become unacceptably high, and government spending and water subsidies came under increased scrutiny. The old paradigm based on the philosophy of larger dams and reservoirs began to fail. People started to become more concerned about environmental loss and economic and social issues. High value had been placed upon the maintenance of the integrity of water resources and the flora, fauna and human societies that have developed around them. During the last three decades of the twentieth century, there have been three important paradigm shifts affecting the water sector: (1) the recognition of water as a basic human right; (2) the acknowledgement of water as an economic good; and (3) the shift from the hard path or supply side policies to the soft path or demand side policies. The following sub-sections discuss the evolution of these three approaches, with particular reference to the various international conventions and statements.

1.2 HUMAN RIGHT TO WATER

Beginning in the 1970s, a series of international environmental or water conferences have taken on the issue of access to basic resource needs and marked the transition towards a more explicit recognition of the right to water. Resolution II on 'Community Water Supply' of the 1977 United Nations Water Conference, held in Mar del Plata in Argentina, recognised water as a right for the first time declaring that *'All peoples, whatever their stage of development and social and economic conditions, have the right to have access to drinking water in quantities and of a quality equal to their basic needs'* (UN, 2011a). The 1979 Convention on the Elimination of Discrimination against Women and the 1989 Convention on the Rights of the Child served as two binding instruments that obliged national governments, as signatories to the conventions, to protect and realise the rights of women and children to water (UN, 2011a).

The challenges of water provision and pricing started to gain increased attention during the 1990s, when the focus of the international community shifted to sustainable development with greater emphasis on the economic and environmental aspects of growth. In 1992, the International Conference on Water and Sustainable Development, held in Dublin, Ireland, recognised the increasing scarcity of water. Principle 4 of the Dublin Statement stated that:

> *"Water has an economic value in all its competing uses and should be recognised as an economic good. Within this principle, it is vital to recognise first the basic right of all human beings to have access to clean water and sanitation at an affordable price. Past failure to recognise the economic value of water has led to wasteful and environmentally damaging uses of the resource.*

Managing water as an economic good is an important way of achieving efficient and equitable use, and of encouraging conservation and protection of water resources." (ICWE, 1992)

In the same year, Chapter 18 of Agenda 21, an outcome of the United Nations Conference on Environment and Development (UNCED), held in Rio de Janeiro, endorsed the resolution of the Mar del Plata Water Conference and called this 'the commonly agreed premise' (UN, 1992). These statements, followed by others, gradually affirmed the roles of the national governments and international organisations, to take responsibility to fulfil the stated 'right to water'. The 1999 UN General Assembly Resolution (A/RES/54/175) 'The Right to Development' stated in Article 12 that '*in the full realisation of the right to development, inter alia: (a) The rights to food and clean water are fundamental human rights and their promotion constitutes a moral imperative both for national Governments and for the international community*' (UN, 2000a). In 2002, the General Comment 15 of International Covenant on Economic, Social and Cultural Rights (ICESCR) provided guidelines for the interpretation of the right to water, framing it within two articles, Article 11, the right to an adequate standard of living, and Article 12, the right to the highest attainable standard of health. In Paragraph 1 it stated that '*The human right to water is indispensable for leading a life in human dignity. It is a pre-requisite for the realisation of other human rights,*' and in Paragraph 2 it mentioned that '*The human right to water entitles everyone to sufficient, safe, acceptable, physically accessible and affordable water for personal and domestic uses*' (UN-OHCHR, 2002).

The various international declarations made after the Dublin Statement and UNCED confirm that water has an economic value and when pricing water, factors such as public right to access to water, the need to maintain affordability, and reduction of wastage, needs to be taken into consideration this. For example, the Ministerial Declaration of the 2nd World Water Forum organised by the World Water Council, held in Hague in 2000, declared *inter alia* that:

> *"To manage water in a way that reflects its economic, social, environmental and cultural values for all its uses, and to move towards pricing water services to reflect the cost of their provision. This approach should take account of the need for equity and the basic needs of the poor and the vulnerable"* (World Water Council, 2000).

Ministerial Declaration of the 3rd World Water Forum, held in Kyoto in 2003, declared *inter alia* that:

> *"Funds should be raised by adopting cost recovery approaches which suit local climatic, environmental and social conditions and the 'polluter-pays' principle, with due consideration to the poor. All sources of financing, both public and private, national and international, must be mobilised and used in the most efficient and effective way."* (World Water Council, 2003)

All the conventions, conferences, declarations and action plans, discussed so far, aimed to address the problems ensuring safe water supply to all. The most notable of these was the United Nations Millennium Declaration 2000, which addressed eight Millennium Development Goals (MDGs) that all 193 United Nations member states and at least 23 international organisations have agreed to achieve by the year 2015. Target 7C under Goal 7 aims to halve, by 2015, the proportion of the population without sustainable access to safe drinking water and basic sanitation (UN, 2000b). Moreover, to emphasise the urgency of dealing with water security issues, the UN General Assembly issued a number of resolutions. In 1992, it declared March 22 of each year as the 'World Water Day', in 2000 it proclaimed the year 2003 as the 'International Year of Freshwater' and in 2003 it decided to mark 2005–2015 as the 'International Decade for Action – Water for Life'. The goal of the decade has been to provide greater focus on water-related programs and projects to achieve the internationally-agreed, water-related goals. Finally, in July 2010, the UN General Assembly adopted a resolution (A/RES/64/292) that *'recognised the right to safe and clean drinking water and sanitation as a human right that is essential for the full enjoyment of life and all human rights'* (UN, 2010). Following this resolution, in September 2010, The Human Rights Council took a further step by specifying that the right to water and sanitation is legally binding.

Although the 'right to water' has been implicitly and explicitly reiterated in a series of international documents, the actual scope, content and obligations, mentioned in these have been subject to different interpretations over the years. Some of the common misconceptions with respect to the right to water are that water should be provided free of cost, that people are entitled to individual household connections and are allowed unlimited use of water (UN, 2011b). The report of the United Nations High Commissioner for Human Rights (UN, 2007) have elaborated the scope and content of obligations to the human right to water and clarified ideological notions that are usually linked with it (Table 1.1).

According to the ICESCR, with respect to the human right to water, State Parties have three major obligations: to respect, to protect and to fulfil the right. The obligation to 'respect' requires governments to refrain from engaging in any activity that limits access to safe drinking water and must also guarantee that individuals have access to effective judicial remedies. The obligation to 'protect' requires the government to adopt necessary and effective legislative and other measures to prevent third parties from denying access to safe drinking water. It is a misconception that governments who fail to achieve universal access are violating the 'right to water'; under the ICESCR, the obligation to 'fulfil' requires the states to adopt legislative, administrative, policies, programmes and others necessary measures to 'facilitate and promote' universal access to safe drinking water and sanitation (UN, 2007).

Table 1.1 Interpretation of terms with respect to right to water.

Term	Interpretation
Sufficient (Quantity)	The water supply and sanitation facility for each person must be continuous and sufficient for personal and domestic uses, which include drinking, personal sanitation, washing of clothes, food preparation and personal and household hygiene. According to the World Health Organisation (WHO), 50–100 litres of water per person per day are needed to ensure that these basic needs are met.
Safe (Quality)	The water required for personal or domestic use must be free from micro-organisms, chemical substances and radiological hazards that constitute a threat to health. Measures of drinking-water safety are usually defined by national and/or local standards. WHO's Guidelines for drinking-water quality provide a basis for the development of national standards.
Acceptable	Water should be of an acceptable colour, odour and taste for personal or domestic use. All water and sanitation facilities and services must be culturally appropriate and sensitive to gender, lifecycle and privacy requirements.
Physically accessible	According to WHO, the water source has to be within 1 km of the home and collection time should not exceed 30 minutes.
Affordable	The costs for water and sanitation services should not exceed 5% of a household's income, meaning services must not affect peoples' capacity to acquire other essential goods and services, including food, housing, health services and education.

Source: UN (2007).

1.3 WATER AS AN ECONOMIC GOOD

In 1992, the Dublin Principles claimed 'water as an economic good' for the first time in a UN setting. However, commercialisation of water supplies through participation of the private sector has been practiced in Europe and the United States as early as the late eighteenth century. During that period, rapid urbanisation in North American and European cities created demands for water networks, mainly driven by the need for fire-fighting and public health concerns (Prasad, 2007; Bakker, 2010, p. 83). In New York in 1795, poor quality of water sold by vendors and filthy public wells led to a major epidemic of yellow fever, which pressurised the city authorities to seek private sector investment for providing clean drinking water. However, poor performance and corruption by the private operator, the Manhattan Company, and series of mishaps (large fire and cholera outbreaks) between 1828 and 1835, forced the government to take over the water supply system (Prasad, 2007). Private companies also supplied water to Boston, Philadelphia, New Orleans and other big cities in USA till the mid nineteenth century. However, most of them were eventually returned to municipal ownership as private operators, driven by the profit motive, were reluctant to increase coverage outside wealthier communities and also failed to provide water for cleaning streets, public fountains and fire hydrants (Hanson, 1959). Hence, compared to 93.8% of waterworks being owned by the private sector in 1800 in USA, the proportion of privately owned waterworks fell to about 30% (2950 out of 9850 waterworks) in 1924 (Masten, 2004).

Similarly, by the beginning of the nineteenth century, private water companies had already been serving the city of London for more than two centuries (Gentry and Auyuan, 2000). However, after the major cholera outbreak in 1840, these became regulated entities and over time, the provision of water and sewerage services in England came to be more dominated by the public sector. In 1861, the private sector served about 60% of the population in larger towns and the share decreased to 20% in 1881 and 10% in 1901 (Prasad, 2007). During the period from 1900–1974, municipalities were mostly in charge of water supply. There were 10 government-owned regional water authorities supplying water and sewerage services and 29 statutory water companies supplying water only. Following the introduction of the Water Act in 1989, Britain privatised its 10 regional water and sewerage authorities by selling its assets and liabilities (OFWAT, 1993).

In France, private sector participation in the operations of water systems started in 1782 when a company founded by the brothers Perrier was granted a license to supply piped water in an area in Paris for 15 years (Roth, 1987). This was followed by the first municipal contract won by Generale des Eaux (currently Veolia) in 1853 under the reign of Napoleon III (Financial Times, 1999). Unlike Britain which opted for complete divestiture, France used the 'delegated management' model, under which municipalities retain ownership of infrastructure and the right to impose tariffs while private companies bid for long-term contracts to provide the services (Financial Times, 1999).

The post Second World War period was characterised by the notion of the 'development state' during which governments where considered as 'prime movers' of economic development. The central philosophy was that rapid economic growth can only be fostered by diverting factors of production from primary sectors typified by traditional technology, low productivity and decreasing returns to industrial sectors that would yield higher productivity and increasing returns (Adelman & Yeldan, 1999). Hence, the 'big push' strategy, based on the assumption that developing countries could achieve rapid economic development by following the path taken by the West, was adopted. Under this model, development organisations and governments focused on financing large scale infrastructure to stimulate production (Bakker, 2010, p. 57). During this time, the World Bank, as the largest multilateral lender, played a leadership role in shaping water policy and lending, and targeted investment towards large water dams that could support agricultural modernisation, urbanisation and industrialisation. This move away from private sector towards greater state involvement was justified by the supposed existence of 'market failures', such as natural monopoly and environmental externalities, along with the reluctance of the private sector to invest in these highly capital intensive and less profitable projects. Moreover, it was expected that governments could use cross-subsidies to expand water services to the poorer customers (Bakker, 2010, p. 85).

Although large sums of international aid and multilateral loans were given since 1950s, public authorities only focused on central affluent urban areas, leaving the peripheral and rural areas underserved (Winpenny, 2003). Soon these organisations concluded that governments acting as both owners and regulators of water utilities were prone to the 'poacher-gamekeeper' problem leading to deteriorating performance, corruption and inefficiency due to lack of incentives (Bakker, 2010, p. 43). These 'state failures' were accompanied by the public sector's indebtedness and lack of access to commercial finance. By the end of the 1980s, a decade designated the International Drinking Water and Sanitation Decade, international actors in the water sector started to reach a new consensus that private sector participation could address the problems of water supply and sanitation in the developing countries (Budds & McGranahan, 2003). During this period, the post-World War II view of the government as the 'prime mover' in economic development shifted to the perception of the government as 'the problem' or cause of underdevelopment (Adelman & Yeldan, 1999). Using their leverage as creditors, the World Bank and the International Monetary Fund, aggressively promoted neo-liberal reforms to indebted governments of developing countries through structural adjustment policies that advocated reduction of state spending. These reforms, termed as the 'Washington Consensus' in 1989, emphasised that economic development should be undertaken by the private sector within free markets, with the state playing a facilitating and regulatory role without direct involvement (Budds & McGranahan, 2003). Moreover, the official development assistance for water supply and sanitation started to decline in the 1990s, preceded by three decades of steady increase (Bakker, 2010, p. 73).

After the 1992 Dublin Statement, the 'water as an economic good' principle has been used strongly to justify commercialisation of water supply and to promote full cost recovery although it contradicted the notion of human rights to water (Prasad, 2007). Over the next two decades, many developing countries have embarked on conducting water sector reforms which have been usually emphasised on the economic dimensions of water and a greater role of the private sector, raising concerns of equity and sustainability and a fundamental conflict between an economic approach to water and a rights based one. While a rights based approach entails provision of safe drinking water to all at affordable prices, an economic approach requires treating water as a commodity that should be priced at cost-effective levels. Although these basic perspectives tend to fuel a 'commodity versus rights' debate, in reality, the two approaches are mutually dependent and one can rarely be achieved without addressing the other. This is because, although water as a natural resource is free of cost and individuals have rights to acquire it for example, by digging wells or collecting it from rivers/ponds, provision of clean and treated piped water requires heavy investments in infrastructure and operations and maintenance. If water is not adequately priced, utilities have to depend on government subsidies and donor assistance, and in cases where such funds are limited, utilities suffer from poor financial health that ultimately translates to poor quality of service.

Treating water as an economic good does not necessarily mean that water prices will be set by the market forces of demand and supply. In almost all countries, water prices are fixed and regulated by public authorities, and in cases where tariffs are relatively higher, the poor can be supported by means of different social policies, such as cross-subsidies, increasing block tariffs (IBT), lifeline consumption and ex-post assistance. The 'commodity versus rights' debate has also been used by scholars, policy makers and businessmen to support their ideological notions with regard to public versus private sector provision of water supply services. The arguments most often mobilised by proponents of the public sector are based on the innate characteristics of water being a public good which leads to a natural monopoly and is linked to principles of basic human rights (Budds & McGranahan, 2003). On the other hand, supporters of privatisation argue that water is an economic good subject to full cost pricing and that the private sector can mobilise funds and result in greater efficiency due to increased competition. Here it should be noted that while treating water as a human right and public good does not necessarily entail public sector provision, recognising its economic dimensions does not also require the engagement of the private sector.

1.4 WATER USE EFFICIENCY

Today about half of the world's population live in urban areas and the trend shows that urban population in low and middle income countries will continue to increase in years to come (Figure 1.1). The increased urban population is putting immense pressure on basic goods and services like housing, energy supply, sanitation and sewerage facilities, water supply and others. The cost of providing water supplies to urban centres has steadily increased over the years and will continue to increase, as nearby sources are exploited and it becomes essential to transport water from far afield or extract it from great depths (see Box 1.1). Reliance on shared water resources or import of water from neighbouring countries or states can lead to conflicts and uncertainty in resource availability.

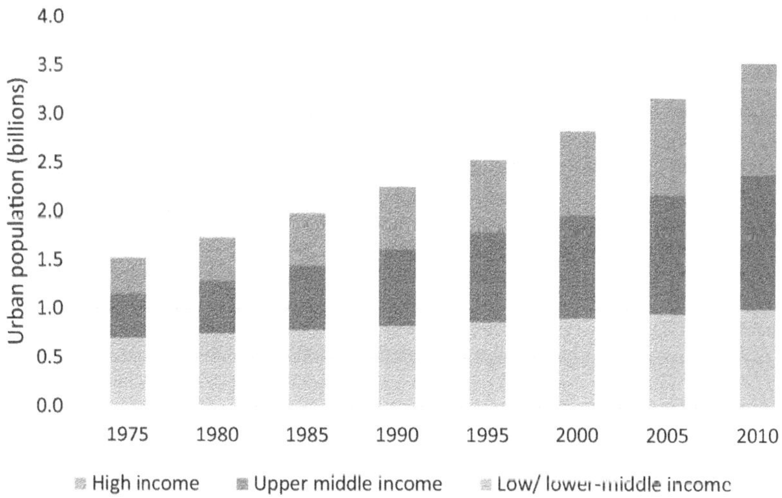

Figure 1.1 World's urban population (Data from World Bank, 2012).

BOX 1.1 WATER SUPPLY AND DEMAND SITUATION IN BANGALORE, INDIA

Driven by job opportunities and economic growth brought about by the IT sector, Bangalore has experienced a rapid population growth and expansion of its urban boundaries in the past decade. In 2007, the Bruhat Bangalore Mahanagara Palike (BBMP), or the Greater Bangalore City Corporation, was created by merging the existing corporation with seven city municipal councils (CMCs), one town municipal council (TMC), and 110 villages surrounding the city, bringing the total area up from 225 to 740 km^2. The estimated population of BBMP's 198 wards as per the 2011 census is 8.5 million, up from 4.6 million in 2001. This population growth has mostly occurred in the fringes, where poor migrant construction workers live in 'slum-like' conditions in non-durable housing structures, often interspersed with middle class houses. The increased population poses tremendous pressure on the city's water supply services. As of January 2013, BWSSB has about 0.66 million connections while at least 1.6 million properties are estimated to be within the BBMP service area (Navya, 2013).

The Bangalore Water Supply and Sewerage Board (BWSSB) sources most of its water from the Cauvery river with a very small proportion from Arkavathy river. The development of infrastructure required to abstract and transport water from the Cauvery river has been carried out in four stages since 1969, with Cauvery Stage IV Phase II being completed in 2012. Although the latter was supposed to add an additional 500 million litres per day (MLD) to BWSSB's existing supply of 900 MLD, only 220 MLD would flow initially (Balasubramanian, 2013). With the end of this project, BWSSB has already utilised the maximum water allocated for the entire

(Continued)

BOX 1.1 (*Continued*)

urban and rural areas in the Cauvery basin within Karnataka. According to the verdict given by Cauvery Water dispute Tribunal in 2007, 270 billion cubic feet water has been allocated annually for the state of Karnataka, of which 17.22 billion cubic feet (equivalent to 1400 MLD) has been earmarked for urban areas and 8.7 billion cubic feet to Bangalore. Since half of Bangalore lies outside the Cauvery basin, these areas are not entitled to water drawn from the river.

As the river is 100 km away from the city and as Bangalore is situated at a height of 1000 m above sea level, a lot of electricity, amounting to 75% of the board's expenditures, is used to pump the water. For treatment, conveyance and distribution, BWSSB incurs a cost of Rs.16 per m^3 of water. However, domestic consumers pay a unit charge of Rs.6 for the first 8 m^3, Rs.8 for the second block of 9–25 m^3 and Rs.15 for the third block of 26–50 m^3 per month (BWSSB, 2013a).

In order to connect people living on the outskirts of Bangalore to piped water, the Greater Bangalore Water and Sanitation Project (GBWASP) was initiated in 1998 to distribute Cauvery River water by laying new distribution pipelines in the periphery. After initial abandonment, the project was resumed in 2003 when the state government recruited FIRE-D to develop a market-based financing framework for GBWASP. In order to make the project 'bankable', FIRE – D designed an innovative model of 'pooled finance' in which capital for the project was to be collected through Beneficiary Capital Contributions (BCC), state loans, grants and debt raised through municipal bonds.

After couple of revisions, the final BCC scheme had been designed such that poor households, classified as those living in dwelling size of less than 600 sft, had been given a full waiver, while residential plots of 600–1200 sft, 1200–2400 sft and above 2400 sft have to pay Rs. 5000, Rs. 10,000 and Rs. 15,000 respectively (BWSSB, 2013b). From an equity standpoint, charging peripheral customers for capital costs may not seem justified because the core customers, who have never had to incur any such costs, are receiving water from the same source.

The water supply from Cauvery Stage IV has been entirely allocated for the seven CMCs and Kengeri TMC and since 2005 pipelines have been laid in these areas as part of GBWASP. However, no water has been supplied through most of these pipes till date. According to BWSSB engineers, unless all households have applied for GBWASP connections and have paid upfront as per the BCC scheme, water cannot be supplied to even those who have taken connections. This is because, the individual pipes do not have valves installed and unless all households are connected water will spill out the open ends and flood the roads. As of January 2013, although 140,000 households have been given GBWASP connection till their doorsteps, only 40,000–50,000 have paid for the connections and installed individual plumping fixtures inside their premises (Balasubramanian, 2013).

Households who have not yet paid upfront for the GBWASP connections are often reluctant to do so because BWSSB have not yet supplied Cauvery water to many of those who have already paid. In 2012, BWSSB launched a simplified application procedure, named 'Sajala' for new water connections. Under this scheme, households who have not paid the BCC charges during 2005–07, will have to pay additional fines of Rs. 3000, Rs. 6000 and Rs. 9000 for residential plots of 600–1200 square feet (sft), 1200–2400 sft and above 2400 sft respectively (BWSSB, 2013b). By the end of 2012,

(*Continued*)

BOX 1.1 (*Continued*)

residents of some GBWASP areas could sporadically see water gushing out from the blue pipes sticking out of the pavements, when BWSSB tested the system and flushed out years of dust.

Apart from paying upfront costs for a project that failed to meet its deadline for water supply, households also have to pay daily coping costs for purchasing water from private vendors and neighbours. The situation is even worse for poor households living in both 'declared' and 'undeclared' slums within the city. According to research conducted by Gronwall (2013) in six wards in Bangalore, most households use a 'combinator strategy' and obtain water from various sources to meet their daily needs. In these six wards, 49% of poor households used public standpipes, 23% used borewells or open wells, 38% had access to piped water (including both Cauvery water and reticulated groundwater network), 25% purchased water from private tankers and 14% used other sources (fetched water from neighbours or purchased 20 litre packaged water jars) as their main source of water for domestic use (Gronwall, 2013). The volume of water used is less than 50 litres per capita per day (lpcd), which is much lower than the 150 lpcd average used for demand calculations in urban areas. The cost of water from private tankers (with capacities ranging from 5000–9000 litres) is sometimes included in the house rent, shared by groups of households paying Rs.350–500 per tanker, or free when provided by local politicians to secure votes (Gronwall, 2013).

Due to the poor surface water supply situation in Bangalore, the dependence on groundwater has increased steeply over the years. In 2010, the Karnataka State Department of Mines and Geology estimated that there are about 225,000 borewells in the city (Department of Mines and Geology, 2011). Of these 105,000 well-owners are registered with BWSSB (paying Rs.50 per month) and 13,000 wells have been drilled by the utility itself. Under the Karnataka Groundwater (Regulation and Control of Development and Management) Act (2011), borewell registration have been made compulsory to check further exploitation of groundwater and to keep tabs on those using groundwater for commercial purposes (The Hindu, March 2013). Although there are penalties for not registering within the stipulated time, the response has been very poor, partly due to lack of awareness. Together these borewells extract 300 MLD, which is 3.7 times greater than the recharge from city's annual rainfall of 900 mm (Balasubramanian, 2013). In many areas, borewells have gone deeper than 1000 feet and many have dried up as well.

In order to develop alternative sources of water supply, in 2011 the state government amended the Bangalore Water Supply and Sewerage Act (Article 72A) and incorporated mandatory rainwater harvesting provisions. According to the law, every owner or occupier of a building having site area greater than 2400 sft (222 m^2) or every owner proposing to construct a building on a site area greater than 1200 sft is required to construct rain water harvesting structure for storage for use or for ground water recharge. Water experts in the city have derided the initiative, mentioning that the act only covers about 44,000 houses out of 1.6 million properties (Balasubramanian, 2013).

Many solutions have been suggested by experts to ease the water supply problem in Bangalore. One of the essential steps is to reduce the high percentage of unaccounted

(*Continued*)

BOX 1.1 (*Continued*)

for water (UfW) (about 45%) by reducing leakage from distribution pipes. Although replacing pipes in a city as large as Bangalore requires significant funds and time, the process can be initiated using management contracts such as those in Hubli–Dharwad, Gulbarga and Belgaum. The depleting groundwater aquifers can be recharged at city or household level using recharge pits and the city lakes should be cleaned and managed properly. Separation of sewage discharge from storm water drainage pipes and rainwater harvesting at city-scale (like Singapore) may seem more farfetched.

Note: USD 1 = Rs. 46.7 (2011 average exchange rate from World Bank, 2012).

As discussed above, cities have traditionally followed the 'hard path' to expand the freshwater supply for its population by constructing more centralised infrastructure related to water extraction, treatment and distribution. However, well-known water authors such as Postel (1998, 1999, 2001) and Gleick (2000b, 2002, 2003) have questioned the sustainability of the current system if it is not transformed in a significant way. Gleick (2002, 2003) prescribes the 'soft path' – a term coined by energy analyst Lovins (1976) – where the solution to water scarcity and increasing demand lies in increasing 'water productivity' rather than endlessly seeking new supplies. Although not often discussed, an interesting analogy lies between the energy sector and water sector analyses and debates. For example: While Lovins argues that the alternative energy future lies in increasing energy efficiency and diversifying energy production methods with special reliance on 'soft energy technologies' like solar, wind and so on, Gleick mentions the future of freshwater supply also depends on demand side management techniques such as enhancing water conservation using 'water efficient fixtures' and other price and non-price tools.

Conservation of water resources is one of the important aspects in ensuring sustainable development of cities and should incorporate environmental, social as well as economic dimensions in it. Chapter 18 of Section II of Agenda 21[1], titled 'Protection of the Quality and Supply of Freshwater Resources', highlights the importance of equitable access to adequate freshwater as a vital component of life, while preserving the earth's ecosystems. Under its recommended activities, it specifically encourages countries to '*promote schemes for rational water use through public awareness-raising, educational programmes and levying of water tariffs and other economic instruments*', as well as '*promote water conservation through improved water-use efficiency and wastage minimisation schemes for all users, including the development of water-saving devices*' (UN, 1992).

According to Baumann *et al.* (1984), any water management practice is said to qualify as conservation if it conserves a given supply of water through reduction of water use or water loss as well as result in a net increase in social welfare, that is, the benefits of resource saving exceeds the costs incurred. Hence, water conservation can be defined as 'any beneficial reduction in water use or in water loss' (Baumann *et al.*, 1984). In this study, water conservation strategies have been categorised as non-price and price strategies. Non-price strategies include outreach and education to raise public awareness about the importance in conserving water and policy and regulation to establish water use requirements and restrictions. Price strategies include the different forms of water and wastewater tariffs such as increasing block rates, water budget based rates and seasonal rate structures.

Although the concept of water conservation has been reiterated in international documents, national water policies and academic literature, the implementation of water saving measures in urban households is still largely limited to the developed world. This is because challenges faced in developing and maintaining a sustainable urban water supply system are different in the urban areas of developing and developed countries. Water agencies and utilities in developing country cities often are plagued with institutional and financial problems such as lack of funds for capital investment and maintenance, shortage of tariff revenues for operating expenses, high levels of non-revenue water, illegal connections and corruption, inefficient tariff rates, lack of universal coverage, poor quality of water and absence of continuous supply. Under such circumstances, expanding coverage, ensuring reliable water supply in terms of quality and quantity and securing enough funds for recurrent expenses are priorities, and reliance on donor funded projects continue as the only mechanism for developing new infrastructure. Other aspects of an environmentally sustainable system become difficult to achieve and hence, notions of water conservation, use of recycled wastewater, proper treatment of sewage prior to discharge into water bodies, development of non-conventional sources and promotion of public awareness are rarely addressed.

Environmental sustainability and economic efficiency are usually concerns of mature water supply systems, which are found in developed cities, where most water purveyors have sound institutional and financial capacities and almost all households have access to continuous supplies of potable water. Although water utilities in developed cities also are often subsidised by their governments, their tariff revenues are usually sufficient to meet operating expenses and to generate net revenue. The challenge is to maintain the financial and environmental sustainability of the system in the face of growing demands and natural resource constraints.

This general difference in assessment of urban environmental challenges has been theorised by the 'Urban Environmental Transition', developed by Marcotullio and McGranahan (2007). The framework works on the hypothesis that cities environmental burden varies in quality, impact, spatial and temporal scales, depending on their level of development. As cities move from developing to developed status, the environmental burdens shift from 'brown' to 'grey' to 'green'.

Brown issues are those relating to basic urban problems such as sanitation, access to clean water, waste management and slums; 'grey' problems which are problems related to industrial development and infrastructural provision which are mostly locally bound but have wider spatial impacts, such as pollution; while 'green' issues in contemporary cities are usually identified as those relating to climate change management and energy use which may have less immediate impact but are very crucial in the long run, and could have wider impact on global scale. While the concerns discussed above reflect the general situations, such dichotomy may not always exist, as good governance and local resource scarcity may also drive utilities in developing countries to address cost recovery, water conservation and wastewater recycling.

1.5 THE BENEFITS OF WATER CONSERVATION

Water use efficiency in the domestic sector has huge potential to reduce the water demand in a city and has numerous of benefits associated with it. These benefits not only accrue to the end users, that it, the bill paying consumers, but to the society at large. The water utilities enjoy lower operation and maintenance costs, deferred facilities expansion and lower need to purchase water from whole sellers; while the environment benefits from decreased extraction of freshwater and lower discharge of wastewater.

1.5.1 Benefits to consumers

For consumers, the main economic incentive for conserving water is reduction in monthly water bills as well as energy bills associated with heating the water. However, such economic incentive may not be much significant in cities where water and wastewater tariffs are low and comprise of a very proportion of the household's monthly expenditures. The efficacy of different types of rate structures in achieving water conservation has been discussed in Chapter 3.

1.5.2 Benefits to utilities

Reduction in water demand through conservation will lower pumping energy expended to acquire, treat, and distribute the water as well as collect, treat and dispose the wastewater generated. While this will reduce energy costs of water utilities, it can have secondary benefits of reduced energy production by power companies. The volumes of chemicals, such as chlorine, used in water and wastewater treatment are reduced.

Lowering the rate of increase in demand can postpone the construction of water supply and wastewater treatment facilities or avoid them altogether. The types of water utility capital facilities most likely affected include water storage reservoirs, raw-water transmission facilities, water treatment plants, finished water storage, and groundwater pumping stations. Fewer or smaller facilities also reduce staffing costs. Reduction in average day demand affects how much water must be developed, or imported and stored, prior to treatment and use. Reduction in peak day demand affects the sizing and timing of water treatment plant expansions and treated water storage. Water pipelines and pumping stations are affected by peak hour pumping. For example: The installation of 2400 residential water meters in 1991, accompanied by an intensive water conservation programme, costing a total of CAD 550,000 (USD 555,555), enabled the town of Port Elgin, Ontario, to avoid a CAD 5.5 million expansion of its water treatment plant. In 1993, this lead to 25% reduction is domestic water use and 30% drop in wastewater flow, allowing the town to save further CAD 12,000 (USD 12121) in water and sewage treatment operating costs (CCME Water Use Efficiency Task Group, 1994).

In cities, where availability of local water resources are not enough to meet the growing demands, water utilities often purchase water from outside the basin. Reduction of local demand can allow these cities to lower their purchase of raw or treated water, thus reducing costs. For example: Since the late 1990s, 70–80% of the freshwater used in Hong Kong is being imported from Dongjiang at Guangdong, a province in southern China. According to the existing agreement, Hong Kong is entitled to an ultimate annual supply quantity of 1100 million m^3, while the actual drawdown is determined by the amount of rainfall within Hong Kong. The price of the imported water was HKD 3146 million (USD 404 million) in 2010 and HKD 3344 million (USD 429 million) in 2011 (Water Supplies Department, 2011). Box 1.2 discusses how Singapore diversified its water sources to reduce its dependence on water imported from Malaysia.

BOX 1.2 THE NEED TO CONSERVE WATER IN SINGAPORE

Located 1°N of the equator, Singapore has a tropical rainforest climate. Owing to its geographical location and maritime exposure, Singapore's climate is characterised by uniform temperature and pressure, high humidity and abundant rainfall. The average annual rainfall is around 2340 mm and the average temperature is between 25 and 31°C. There is no clear-cut wet or dry season and rain is experienced every single month, usually in the afternoons and early evenings. Although Singapore is not a water scarce country in terms of the precipitation it receives, the limited and densely developed island state and absence of natural aquifers and lakes offer limited land area for catching and storing the rainwater (Luan, 2010).

Given the limitation of local surface and groundwater resources, availability of freshwater for domestic and industrial purposes has always been an important concern for Singapore. When Singapore became an independent state in 1965, its main source of water supply was water imported from Johor (Malaysia) and supplemented by water from the local catchments, that is reservoirs. The Singapore

(Continued)

BOX 1.2 *(Continued)*

government recognised that these two sources of water would not be able to ensure a stable and sustainable water supply for the country's growing economy and population. The government thus had to explore other avenues for more sources of water supply (Table 1.2).

Table 1.2 Basic facts on Singapore.

Area (km^2)[1]	714
Population (millions)[1]	5.183 (2011)
Population growth rate (% per year)[1]	2.1 (2011)
National GDP per capita (USD at current prices)[2]	52,536 (2011)
Average household size	3.5 (2011); 3.6 (2001)
Household income (USD per month)[3]	5721 (2011) Lowest 10% – 1,285 Highest 10% – 22,656

Source: [1]Department of Singapore Statistics, (2012). [2]World Bank, (2012). [3]Median monthly household income from work among resident employed households.

Within the government of Singapore, the Ministry of the Environment and Water Resources (MEWR) is in charge of policy setting for water and sanitation. The Public Utilities Board (PUB), a statutory board under the Ministry, is in charge of providing drinking water as well as of sanitation and stormwater drainage. Currently, the total water demand in Singapore is 380 million gallons per day (mgd) (1.43 million m^3 per day), 55% of which comes from the non-domestic sector and the rest from the domestic sector. Since its independence in 1965, through strategic planning and investment in research and technology, PUB has built a robust and diversified supply of water known as the 'Four National Taps'. The water supply comprises:

Local catchment water

With no natural aquifers or lakes, as well as little land to collect rainwater, Singapore's strategy has been to create estuarine reservoirs by damming the major rivers. As of 2012, about two thirds of the country's surface area has been classified as partly protected catchment areas with certain restrictions on land use. Rainwater is collected through a comprehensive network of drains, canals, rivers and stormwater collection ponds before it is channelled to Singapore's 17 reservoirs for storage. Depending on the amount of rainfall, the local catchments can supply about 200–300 mgd (PUB, 2013b).

Imported water

Under the two bilateral agreements signed in 1961 and 1962, Singapore has been importing water from the Johor state in Malaysia through a 1 km pipeline that runs along the Johor – Singapore causeway. Under the 1961 agreement, which expired in 2011, PUB had the exclusive right to draw off and use 86 mgd (390,000 m^3) water from Gunung Pulai Reservoirs as well as the Tebrau and Skudai Rivers, while the 1962

(Continued)

BOX 1.2 (*Continued*)

agreement, which expires in 2061, allows PUB to extract 250 mgd (1,100,000 m^3) from the Johor River (Luan, 2010). Under the Water Agreements, Singapore pays the Johor government RM 0.03 for every 1000 gallons (4546 m^3) of raw water and the Johor government pays Singapore RM 0.50 for every 1000 gallons of treated water, although it costs Singapore RM 2.40 to treat every 1000 gallons of water (MICA, 2003). In addition, a further agreement reached in 1999 entitles Singapore to abstract an unspecified amount of water from the Linggui dam in Malaysia at a price that is higher than under the 1962 agreement.

NEWater

NEWater is high-grade reclaimed water produced from treated used water that is further purified using advanced membrane technologies and ultra-violet disinfection, making it ultra-clean and safe to drink. After completion of the largest and latest NEWater plant in Changi in 2011 together with the expansion of the existing four, NEWater plants can meet 30% of the nation's water needs (up to 115 mgd or 520,000 m^3). NEWater is primarily for non-potable industrial uses and supplied via a separate distribution network to wafer fabrication, electronics and power generation industries for process use and to commercial and institutional buildings for air conditioning cooling purposes. A small percentage of NEWater is also blended with raw water in the reservoir which then goes through treatment at the waterworks before it is supplied to consumers as tap water (PUB, 2013b).

Desalinated water

SingSpring, the country's first desalination plant was opened in Tuas in 2005 under a public-private partnership with Hyflux. The plant produces 30 mgd (140,000 m^3) of water and meets 10% of the country's water needs. In 2011, a second contract was awarded to Hyflux for the construction and operation of a second desalination plant in Tuas. The plant will have a capacity of 70 mgd (320,000 m^3) and is expected to be completed in 2013 (PUB, 2013b).

While Singapore has been immensely successful in reducing its dependence on imported water and developing alternative sources to achieve self-sufficiency, it has also put great emphasis on demand side management since the 1980s. Singapore has restructured its water prices (Box 3.1) launched mandatory water efficiency labelling schemes (Box 5.3) and conducted numerous public awareness programs (Box 4.3) over the past three decades, all of which have been discussed in detail in the following chapters of this book.

1.5.3 Benefits to the environment

When less surface or groundwater is extracted, more water is left in nature, where it can lead to environmental benefits. When the rate of surface water extraction exceeds the flow rate of the water body, the latter can become dry and unable to provide the essential ecosystem services to local residents. Over-exploitation of groundwater has led to declining groundwater tables in many developing cities, ultimately resulting in drying of aquifers and loss of sustainable water sources. Moreover, less water consumption means less generation and discharge of wastewater into the environment and less pollution of receiving water bodies. However, these benefits do not accrue to one individual or the utility itself and the benefits cannot be quantified in monetary terms.

1.5.4 Benefits to the economy

The move to water efficiency will trigger new economic activities for water-related manufacturing and service sectors, encouraging new business opportunities and job creation. Increased efficiency also means lower costs to business, leading to increased competitiveness. For some cities, lower dependence on imported water leads to less international tensions and greater national security.

1.6 WATER CONSUMPTION IN URBAN HOUSEHOLDS

In terms of water supply, households in urban areas are heterogeneous. This is especially true for developing cities, where there is significant difference in the socio-economic status of high-income and low-income settlements. While piped water networks usually exist in the formal parts of large cities, urban slum dwellers use multiple sources for their daily water requirements. This choice set can include water sources as diverse as in-house tap connections, public or private wells, public or (someone else's) private taps, water vendors or resellers, tank trucks, water provided by neighbours or water collected from rivers, streams or lakes. The fact that some households utilise more than one source indicate either that their use of a particular convenient source is rationed (implying that additional water must be taken from an alternative source); or that it is relatively cheap to take some water but not all from a particular source (e.g., the household has limited capacity to haul cheap water from a given source, and obtain the rest more expensively from another source); or that water from different sources are used for different purposes (drinking, bathing, cleaning, etc.). Since access to water supply is limited for the urban poor, the daily per capita consumption is also very low compared to rich and middle class with access to piped water. While water demand of high-income urban households can be higher than 200 lpcd, the water consumption of low-income settlements can be less than 50 lpcd. Hence, when consumption is barely enough to meet daily basic needs, the need for residential water conservation does not apply to these low volume users.

The scenario is more homogeneous in case of developed cities, where almost all urban households are supplied with piped water and water quality and reliability of service are satisfactory. Because of this characteristic, household water demand in developed cities has been extensively studied in particular to provide measures of price elasticities and socio-economic characteristics, mainly income (Nauges & Whittington, 2010). In almost all studies, the water demand function is specified as a single demand equation for water provided at the tap. Such an approach implicitly assumes that there is no substitute available for water. Water quality as well as service reliability are not controlled for, in general (for the main reason that there is not much variation in terms of service quality across distribution units). On the other hand, researches regarding water demand in developing countries have been under the form of contingent valuation studies to derive willingness-to-pay for getting a house connection to a piped water network or improved water services; and hedonic price studies to infer the valuation of a piped connection through observations of house prices (Nauges & Whittington, 2010).

In this book, the discussion of water conservation with reference to the roles of prices, policies and technologies has mainly focused on examples from developed cities. The same strategies are equally applicable for developing cities, provided there is good governance, universal coverage, proper metering and billing systems and adequate implementation of policies and programmes. In order to understand the potential for residential water consumption, it is important to know how much water is used and for what activities. The general water use activities within a household or 'micro-components' of water use include bathing, toilet flushing, cloth washing, dish washing, cooking and food preparation, house cleaning, hand washing and drinking. Due to cultural and socio-economic differences, there may be slight variations in the ways in which water is used for these activities. For example, in developed cities bathing mostly refers to showering or occasional bath-tub use, whereas in developing cities bathing using bucket water is also common. Similarly, residents in developed cities usually use washing machines and/or dishwashers while in developing cities cloth and dish washing is mostly done by hand. Consumption of pre-processed and cleaned food from supermarkets may also reduce water consumption in the kitchen compared to cultures where meat, fish and vegetables are cut and cleaned at home. Analysis of the micro-components of water consumption is also important to identify the activities and fixtures which use the most amount of water and hence have the greatest potential for conservation. For example: If 20–30% of the household water is used for toilet flushing, replacing conventional toilets using 12 litres per flush with dual flush toilets using 6 and 3 litres for full and half flush respectively can save significant amount of water.

There have been numerous studies, where authors have estimated water usage in different household activities. The methodologies for measuring water consumption are classified into three groups. The first one involves the direct measurement of the water used by installing meters to each water appliance. For example, Tokyo's municipality adopted this method and set meters to each tap in 76 households for one year. The shortcomings in this study were the limited number of samples, and the necessity for the use of numerous meters (4–6 meters per household) involving the need for cooperation from each household. Despite all this, however, the method used in Tokyo was able to obtain real and accurate data. The second method for measuring water consumption involves obtaining estimations through interviews. Each micro-component is determined by gathering information on ownership of appliances, frequency of use, and the volume of water per use. The Environmental Agency (UK) used this method and estimated data were used for water demand forecasting. Fukuoka (Japan) and Bangkok (Thailand) also used this method and collected information from 100 and 814 interviews, respectively (Little, 1996). Although the data is an 'estimated' value, it is easy to obtain many samples. The third method for measuring water consumption is to collect time-series data of total residential water consumption and to calculate it for each component by the water flow pattern. The American Water Works Association (AWWA), Osaka (Japan), and the East Anglia Region (UK) adopted this method (Edwards & Martin 1995; Mayer *et al.*, 1999; Yamanishi & Manabe 2002). This method is convenient for both the user and the experimenter, because there is no need to set meters to each tap. The only real problem is the cost.

The Southeast Queensland Residential End Use Study (SEQREUS), conducted during 2010–2011, aimed to quantify and characterise mains water end uses in a sample of 252 residential dwellings in four local authorities (Sunshine Coast, Brisbane, Ipswich and Gold Coast) in Queensland, Australia. A mixed method approach was used, combining high resolution water meters, remote data transfer loggers, household water appliance audits and a self-reported household water use diary (Beal & Stewart, 2011). According to the study, the average household water consumption in the four regions combined was 145.3 lpcd. End use breakdown on a per capita basis indicated that, on average, showers (42.7 lpcd: 29%), taps (27.5 lpcd: 19%) and clothes washers (31 lpcd: 21%) comprised the bulk of the water consumption. Water end use breakdowns varied substantially across (and within) the regions examined. This variation is a reflection of several factors including family size and composition, socio-demographic factors and climate. In all the homes measured, there was water use from the toilet, clothes washer, taps and showers. The remaining end uses analysed (leaks, dishwasher, irrigation and bath tub) were reported in some, but not all, of the homes (Beal *et al.*, 2011).

The Residential End Uses of Water Study (REUWS), conducted by the American Water Works Association Research Foundation (AWWARF) in 1999, was designed to obtain specific water end use data from a sample of 1188 single family households across 12 locations in North America. The study found that the mean per capita indoor daily water use was 69.3 gallons (262 litres) (including leakage), of which the greatest proportion of water was used for toilet flushing (Figure 1.2).

It is difficult to obtain reliable data describing per capita domestic water consumption at city level. Data to be reliable when values appear in the Annual Reports of water agencies, on their websites, or in peer-reviewed journal articles. The data depicted in Figure 1.3 represent domestic water consumption only, and include both indoor and outdoor uses. However, care should be taken regarding how the cities calculate water consumption. For example, in Hong Kong, if the calculation includes only the amount of freshwater consumed, the per capita use is 130 litres per day, whereas if seawater used for flushing is also considered, per capita use becomes 220 litres per day.

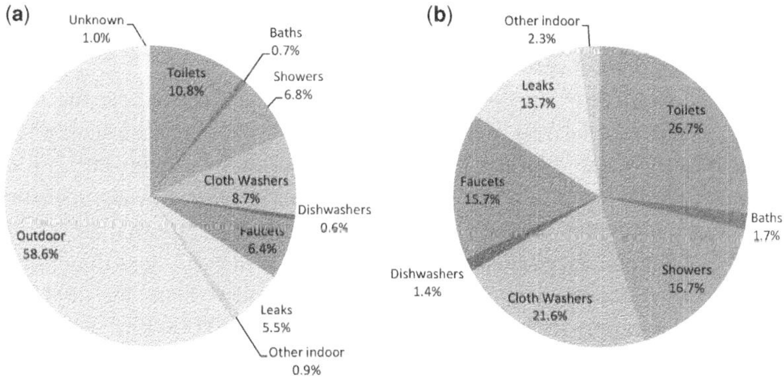

Figure 1.2 Domestic per capita water use in United States (a) both indoor and outdoor; (b) indoor only (Mayer *et al.*, 1999).

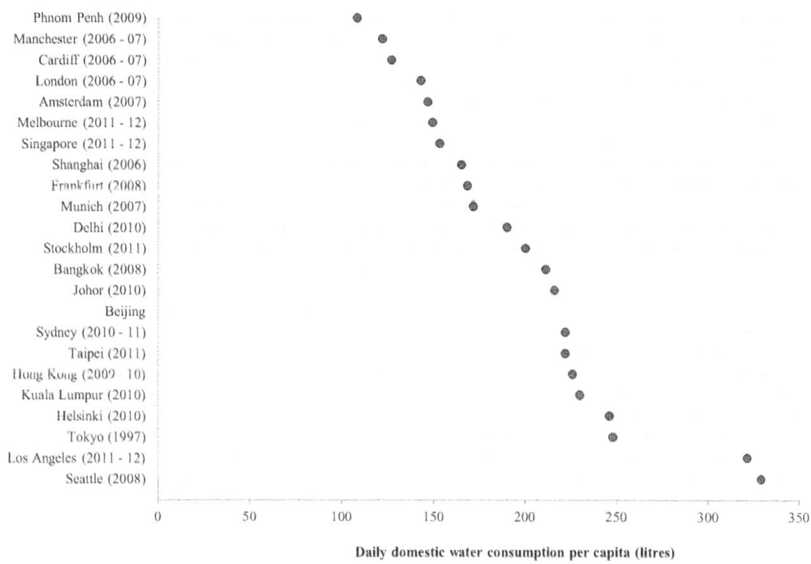

Figure 1.3 Daily domestic water consumption per capita (litres) (Institute of Water Policy, 2013).

Sydney Water reports that the weather corrected per capita water demand in Sydney is 304 litres per day. However, that value has been calculated by dividing the total water supply for households, businesses, industry, and agriculture by the total population served. If one applies the statement that households account for 73% of water consumed in Sydney, the per capita domestic water consumption would be 222 litres per day. The high per capita water consumption in Los Angeles and Seattle is because of inclusion of outdoor water usage.

1.7 CONCLUSION

Exponential population growth, increased urbanisation and rapid economic development are all driving the demand for freshwater resources, the absolute supply of which is relatively constant on a global scale. Shiklomanov (1999) estimated that the total volume of the earth's water is 1.4 billion km^3, of which freshwater accounts for only 35 million km^3, the rest being salt water covering the world's oceans. Moreover, of the 2.5% freshwater available, more than two-thirds is trapped as glaciers and permanent snow cover, leaving only less than one-third in the form of extractable groundwater and surface water. Thus, the total volume usable freshwater supply for humans and ecosystems is 200,000 km^3, which is only 0.01% of all the water on earth (Shiklomanov, 1999; Gleick, 2003).

Moreover, the distribution of these freshwater resources are not uniform across the world, with some countries having abundant supplies while others are facing severe shortages. According to popular Falkenmark Water Stress Indicator (Falkenmark *et al.*, 1989), 'water stress' is said to occur when the annual supply of freshwater is less than 1700 m^3/person, while 'water scarcity' is defined as the situation when the annual supply of freshwater drops below 1000 m^3/person. Countries in North Africa and the Middle-east as well as Asian nations like Pakistan and India fall under the above categories, while many others will be joining the list in next few decades. The Food and Agricultural Organisation (FAO) states that by 2025, 1.8 billion people will be living in countries or regions with absolute water scarcity, and two-thirds of the world population could be under stress conditions (FAO, 2011).

Agriculture is the largest sector of global freshwater use, followed by domestic and industrial sectors; however, the relative proportion of each varies from country to country. Based on data on the total freshwater withdrawal by country and sector from 'The World's Water'[2], it is observed that while in an agrarian country like Bangladesh the proportions of annual water withdrawal for agricultural, domestic and industrial sectors are 96, 3 and 1% respectively, in an industrial economy like Singapore the proportions are 4, 45 and 51% and in a large developed country like the United States they are 41, 13 and 46% respectively.

Although there is huge potential for water conservation in each of the three sectors, this book addresses the importance of using water efficiently in urban households. Chapter 2 of this book focuses on the determinants of water conservation behaviour, including psychological factors such as values, beliefs and attitudes, socio-economic factors such as income, water pricing and policies, environmental factors such as seasonal variations and demographic factors such as household size and age. Chapter 3 addresses the role of water and wastewater tariff structures in achieving the goals of revenue generation, affordability, demand management and equity and the design of conservation oriented rate structures and Chapter 4 discusses the role of policies such as mandatory water restrictions, labelling of water saving devices and promotion of public awareness. The role of water saving devices in providing technological solutions to household water conservation has been discussed in Chapter 5. In relation to the above issues, the book provides several detailed case studies of cities to understand the effectiveness of such demand management tools and the lessons learnt. Overall, the book aims to provide a comprehensive overview of the various price and non-price tools that can be used to manage domestic water consumption.

ENDNOTES

[1]Agenda 21 is an action plan of the United Nations (UN) related to sustainable development and was an outcome of the United Nations Conference on Environment and Development (UNCED) held in Rio de Janeiro, Brazil, in 1992.

[2]Published every two years by Island Press, The World's Water provides both detailed analysis of the most significant trends and events and up-to-date data on water resources and their use.

Chapter 2

Determinants of domestic water consumption: Theories and observations

Residential water consumption in urban areas is determined by a wide range of demographic, socio-economic and contextual factors. The daily per capita domestic water use shows variations among cities (Figure 1.3) and even among different socio-economic classes within the same city. The general population is usually not at all aware of the amount of water they use for different purposes and conservation is often a neglected issue. This chapter is divided into two main sections – Section 2.1 provides the theoretical background of environmental beliefs and pro-environmental behaviours and Section 2.2 discusses eight main determinants of domestic water consumption.

2.1 THEORIES OF ENVIRONMENTAL BELIEFS AND BEHAVIOURS

Environmental beliefs, or worldviews, of the relationship between humans and their natural surroundings have been mentioned as potential predictors of conservation behaviour (Scott & Willits, 1994). Gray (1985) posits environmental beliefs as underlying a system of attitudes and beliefs that determine behaviour toward the environment. Since the 1970s, a number of theoretical frameworks have been developed by authors from different disciplines to understand the conceptions about humanity–nature relationships.

The 'New Environmental Paradigm–Human Exception Paradigm (NEP–HEP)' (Dunlap & Van Liere, 1978), from sociology, is the most widely used instrument to assess general environmental beliefs. The model works with the hypothesis that, impelled by the obvious environmental problems plaguing the planet, people all over the world are changing from an anthropocentric worldview – an Human Exemptionalist Paradigm – to an ecological worldview, more respectful of nature – or to a New Ecological Paradigm.

Stemming from anthropology, the 'Cultural Theory' (Douglas & Wildavsky, 1983) posits that the bases of contemporary environmentalism lie in deep-rooted orienting dispositions or cultural biases that make some individuals especially fearful of environmental threats to human health and safety. Dake (1991, 1992), following Douglas and Wildavsky, have developed scales that measure four orienting dispositions: egalitarianism, hierarchy, individualism and fatalism. The theory suggests that egalitarians will be most concerned with the environment while individualists will be least concerned.

The 'Post-materialist Values' hypothesis, from political science, associates environmental concern with other post-materialist values emergent in affluent post-war societies (Inglehart, 1977). Since most basic needs are satisfied in these affluent societies, people have started to show concern for a series of non-materialist issues, such as environmental protection (Inglehart, 1995).

Theories of altruistic behaviour, from the field of social psychology, presumes that because environmental quality is a public good, altruistic motives are a necessary for an individual to contribute to it in a significant way. The best developed example of this approach builds on Schwartz (1973) and Schwartz (1977)'s 'Moral Norm-Activation Theory of Altruism'. The theory holds that altruistic behaviour occurs in response to personal moral norms that are activated in individuals who believe that particular conditions pose threats to others (awareness of adverse consequences) and that actions they could initiate could avert those consequences (ascription of responsibility to self).

More recently, Stern *et al.* (1999) integrated the concepts of HEP-NEP and the norm activation theory to develop a Value-Belief-Norm (VBN) theory of environmentalism. The theory postulates that acceptance of the NEP is formally antecedent to awareness of consequences in the norm-activation model. The degree of acceptance of the NEP is itself correlated (positively) with biospheric and altruistic values and (negatively) with egoistic values.

Stern (2000) proposed that environmentally significant behaviour may be defined from an intent-oriented and an impact-oriented perspective, respectively. From an intent-oriented perspective, environmentally significant behaviour is defined by the motivation of the actor to undertake pro-environmental actions. Although some behaviours are performed because they are perceived to be environmentally beneficial, they do not necessarily result in a reduction of the actual impact on the environment. On the other hand, an impact-oriented perspective does not focus on the motivation of the actor to perform certain behaviour but defines behaviour by its actual impact on the environment.

Next to a distinction between an impact-oriented and intent-oriented classification of behaviours, one can discriminate between behaviour that directly, and behaviour that indirectly influences environmental qualities (Stern, 2000). Behaviours that are performed in the private sphere of households, such as energy or water conservation, can be considered as direct behaviours as they have direct environmental consequences (Stern, 2000). Other direct behaviours include environmental activism, such as active involvement in environmental organisations and demonstrations. Non-activist behaviours in the public sphere, such as stated approval of environmental regulations or willingness to pay higher taxes for environmental protection, can be categorised as indirect behaviours. Although behaviours with direct environmental impacts have gained the most attention from psychologists and other consumer researchers, behaviours with indirect impacts, such as policy support, may eventually also have large environmental impacts. Based on a review of existing literature that studied causal variables on behaviour in an environmental context, Stern (2000) has categorised the variables that influence behaviour into four main groups: attitudinal factors, external or contextual forces, personal capabilities, and habit or routines.

Attitudinal factors include values, norms, beliefs, and attitudes. The theory of planned behaviour (Fishbein & Ajzen, 1975; Ajzen, 1991) (see Section 2.2.3 below) is the most widely researched model of the relationship between attitudes and behaviour. The theory suggests that the intention to engage in a particular behaviour is a function of three antecedents: the attitude toward the behaviour, social norms, and perceived behavioural control.

External or contextual forces include interpersonal influences (e.g., persuasion, modelling); community expectations; advertising; government regulations; other legal and institutional factors (e.g., contract restrictions on occupants of rental housing); monetary incentives and costs; the physical difficulty of specific actions; capabilities and constraints provided by technology and the built environment (e.g., building design, availability of bicycle paths, solar energy technology); the availability of public policies to support behaviour (e.g., recycling programs); and various features of the broad social, economic, and political context (e.g., the price of oil, the sensitivity of government to public and interest group pressures, interest rates in financial markets)

Personal capabilities include the knowledge and skills required for particular actions (e.g., the skills of a movement organiser for activism, mechanical knowledge for energy-conserving home repairs), the availability of time to act, and general capabilities and resources such as literacy, money, and social status and power. Socio-demographic variables such as age, educational attainment, race, and income may be indicators or proxies for personal capabilities.

Finally, habit or routine is a distinct type of causal variable. Behaviour change often requires breaking old habits and becomes established by creating new ones. Habit, in the form of standard operating procedure, is also a key factor in environmentally significant organisational behaviour.

2.2 DETERMINANTS OF DOMESTIC WATER CONSUMPTION

Like any other pro-environmental actions, such as recycling, energy saving and anti-littering, water conservation behaviour is determined by a number of personal and contextual factors. Taking inspiration from Stern's four broad variables of environmentally significant behaviour, I have classified the determinants of water conservation behaviour into two broad categories: personal factors and contextual factors, and nine sub-categories, which are illustrated in Figure 2.1 and discussed elaborately in this section.

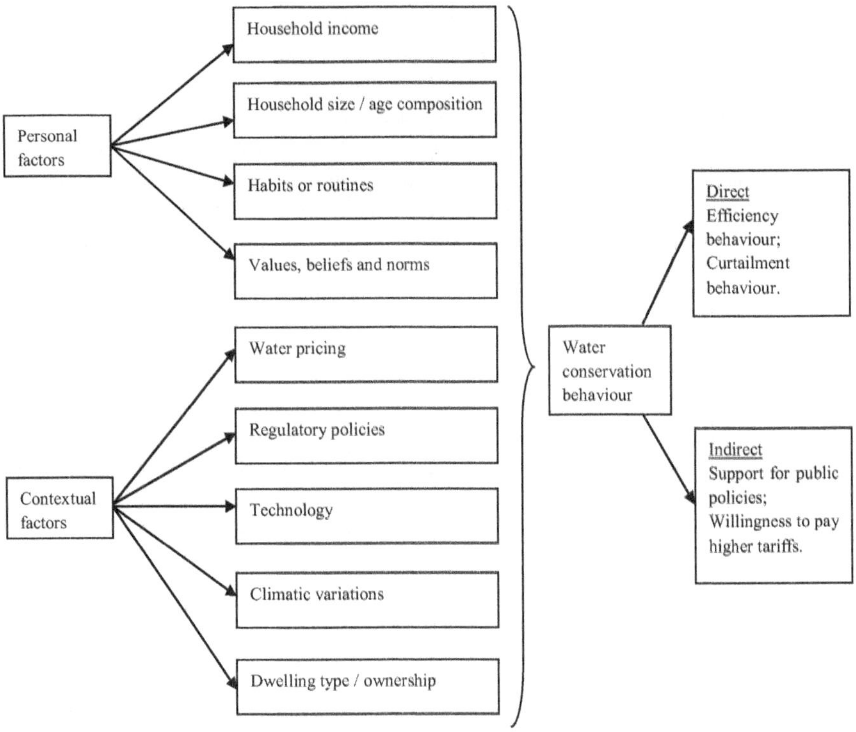

Figure 2.1 Determinants of domestic water consumption.

2.2.1 Household income

The influence of household income on water consumption is a complex issue and the analysis of this relationship varies among authors. According to Mainieri *et al.* (1997) affluence might decrease water use because rich people live in newer homes with more water efficient technologies and also possess environmentally sensitive attitudes due to their better education and awareness. Contrarily, Kurz (2002) mentions that affluence will increase water use because rich people consume more goods and services and environmental attitudes are not always closely linked to income. Rubin (2003) analysed household income and expenditure data of 1.2 million US households collected by the Census Bureau and found that each USD 10,000 increase in annual income, increases water expenditure by USD 15. As the cost of water increases at a much slower rate than annual income, percentage of household income spent on water declines as income increases.

A number of studies have estimated the income elasticity of demand for domestic water consumption. While these studies differ in terms of the study locations, the time periods and data-sets used, they also vary in terms of the estimation techniques, the functional forms and the variable taken into consideration. In basic economic terms, income elasticity is a measure of the responsiveness of demand for a certain good or service due to changes in income of the consumers. Schleich and Hillenbrand (2009) econometrically analysed the impact of several economic, environmental and social determinants on the per capita water demand in Germany based on cross-section data for about 600 water supply areas. Using the Ordinary Least Squares (OLS) and the Instrumental Variable (IV) procedures, the authors estimated that the income elasticity of water demand is 0.355 and 0.457, indicating that water is a normal good and consumption increases less than proportional to higher household income. This is because, in general, households with higher incomes are expected to consume more of the complementary commodities associated with water through having gardens, dishwashers, saunas, or pools, all of which increase indirect water demand. Moreover, Schleich and Hillenbrand (2009) found that the income elasticity is not constant, but decreases with higher income levels, as the percentage share of household expenditure on water decreases. Previous studies (Espey *et al.*, 1997; Arbués *et al.*, 2003; Dalhuisen *et al.*, 2003; Worthington & Hoffman, 2008) have also provided strong empirical evidence that water demand is rather inelastic in terms of income changes. Dalhuisen *et al.* (2003) carried out a meta-analysis of variations in price and income elasticities of residential water demand and tried to link the variation in estimated elasticities to differences in theoretical microeconomic choice approaches, differences in spatial and temporal dynamics, as well as differences in research design of the underlying studies. For instance, the mean and median for the income elasticities surveyed by Dalhuisen *et al.* (2003) are 0.43 and 0.24, respectively, with a standard deviation of 0.79.

In Singapore, it has been observed that the household monthly water consumption increases with income as rich families tend to live in bigger houses, with more water use activities. In 2011, the median household income from work (excluding Central Provident Fund) among resident employed households was SGD 6286 (USD 5110) compared to an average of SGD 8864 (USD 7206) (Department of Statistics Singapore, 2011). Based on the national average monthly domestic water consumption of 20 m^3, the water and wastewater bill is SGD 40.6 (USD 33.84) (excluding Goods and Services Tax of 7%), which amounts to just 0.68% of monthly income. However, a large variation in water use exists among the different dwelling types. Figure 2.2 shows the monthly water consumption for each of the different dwelling types in Singapore and relates this with the median monthly income of each of these categories. Figure 2.3 shows the proportion of each of the different dwelling categories in Singapore. As shown, water consumption in 1 and 2 room Housing Board Development (HDB) flats is substantially lower (11.2 m^3) than the national average, while landed properties, such as terraced houses, semi-detached properties, and bungalows have substantially higher consumption (39.7 m^3). Families living in HDB 1–4 room flats, who constitute 57% of all households, have an average monthly income of SGD 2000 to SGD 5000, while those living in condominiums and landed properties, comprising 11% of all households, earn between SGD 12,000 and SGD 15,000 per month. Thus, not all households pay similar portions of their incomes as water and wastewater bills. Poorer families in HDB 1 and 2 room flats pay as much as 2% of their incomes as water bills, while wealthier families living in condominiums pay as little as 0.5%.

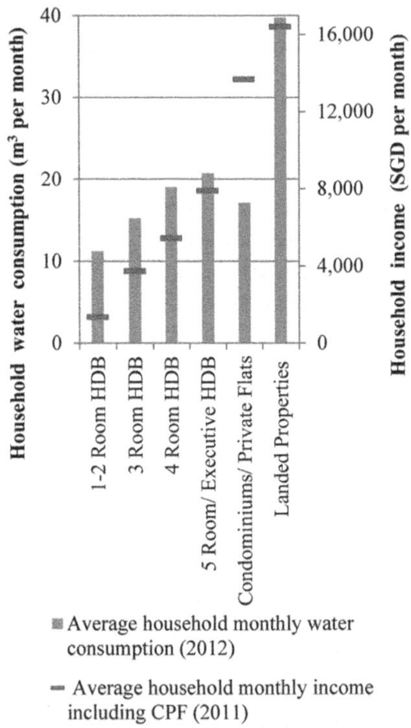

Figure 2.2 Monthly household income (Department of Statistics Singapore, 2011) and water consumption in Singapore (Singapore Power, 2013a).

The Southeast Queensland Residential End Use Study (SEQREUS), conducted during 2010–2011, aimed to quantify and characterise mains water end uses in a sample of 252 residential dwellings in four local authorities (Sunshine Coast, Brisbane, Ipswich and Gold Coast) in Queensland, Australia. A mixed method approach was used, combining high resolution water meters, remote data transfer loggers, household water appliance audits and a self-reported household water use diary (Beal & Stewart, 2011). The results of this study also substantiated that household water consumption increases with income and the main activities associated with the increase usage are showers, clothes washer, dish washer and bath. Households with annual income greater than AUD 90,000 (USD 92,783) consumed about 387 litres per day, whereas those in income categories of AUD 60–90,000, AUD 30–60,000 and less than AUD 30,000 (USD 32,927) consumed 318 litres per day, 325 litres per day and 258 litres per day respectively.

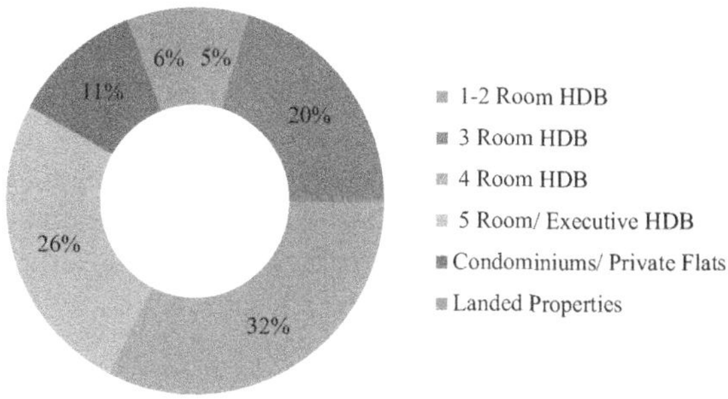

Figure 2.3 Proportions of different dwelling types in Singapore (Department of Statistics Singapore, 2011).

2.2.2 Household size and age composition

In general, a decrease in household size is believed to increase per capita water consumption due to economies of scale. For instance, four people living together are likely to consume less water than four people living separately. Alternatively, Gilg and Barr (2006) found that household size may present a barrier to establishing a water conserving household, either because of the increased difficulty of establishing conservation norms among more people, or the limited physical or financial capacity that may be associated with larger family size.

Martin (1999) studied historical data on household size, population growth and domestic water use patterns for selected countries in the Middle East and concluded that while population growth contributes significantly to increased water use in residential sector, the growth in the number of households and changes in the household size also play a major role. Based on historical and projected data, Martin showed that since 1975 the numbers of households and population have experienced increasing trends and although household size increased from 1975 to 1990s, it is following a decreasing trend in today's modern world. According to Martin (1999), it is likely that decrease in household size will not only increase household water use due to straight-forward economies of scale, but the decrease in household size will also reduce the effectiveness of investments into water saving technical measures. Hence, the combined effect of population growth, increased number of households in urban areas and decrease in household sizes will significantly increase domestic water consumption in coming years.

The SEQREUS, mentioned in Section 1, also found that an increase in the household size resulted in an increase in overall household water use, but a decrease in the per capita water consumption. Moreover, the study also compared the end uses between different household typologies. Households were grouped into single (1 person), adult household (2 people), small family (e.g., 2 adults and 1 child), medium family (e.g., 2 adults and 2 children) and large family (5 or more people). The results again showed that larger families are typically more water efficient on a per capita basis (Beal & Stewart, 2011). The water consumption pattern for a single household shows a relatively even consumption across all end uses with a growing trend for higher clothes washer, shower and tap use as the households become larger. These findings suggest that targeting water conservation programs for specific end uses for single families is likely to not be very effective at reducing total consumption, compared to targeting high end uses such as the shower and clothes washer within larger families. Greater savings would be realised by installing efficient water stock devices in the medium to large family cohorts (Beal & Stewart, 2011) (Figure 2.4).

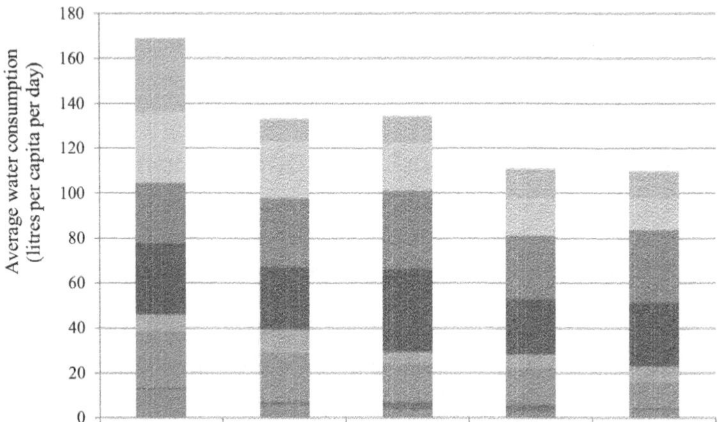

	Single family	Adult family	Small family	Medium family	Large family
Leak	32.5	9.9	11.8	13.2	12.1
Toilet	31.6	25.2	21.3	16.4	13.8
Clothes washer	26.9	30.9	35	28.2	32.7
Shower	32	27.6	36.5	24.6	28.1
Dishwasher	7	10.2	5.6	5.9	6.9
Tap	25.5	22	17.1	16.7	11.8
Bathtub	0.6	1.4	3.5	3	0.8
Irrigation	12.7	5.6	3.4	2.7	3.5

Figure 2.4 Water consumption in southeast Queensland according to household size (Beal & Stewart, 2011).

Interestingly, household water consumption also depends on the age of the occupants and usually younger families with children are associated with more water use than older families with retirees. The SEQREUS study found that the presence of one or more teenagers in the household significantly increased the per capita water used for showers from a mean value of 36.5 to 46.6 lpcd (Beal & Stewart, 2011). Similarly, the Residential End Uses of Water Study concluded that children and teens used incrementally more water for showers and baths than did adults. Both studies mentioned that use of cloth washers increases incrementally with the number of teens living in the household and the number of persons working full-time outside the home. On the other hand, older families had higher use of taps and toilets, mainly because these people spent greater time at home (Beal & Stewart, 2011). The amount of water used for toilet flushing is negatively related to the number of persons employed full-time outside of the home. For those employed outside the home, some flushing at home is replaced by flushing at work (Mayer *et al.*, 1999).

The UK Sustainable Shower Study by Unilever UK and Ireland recorded accurate water consumption data for 2600 showers in 100 British households. The study found that while 12 year old girls spend less time in the shower (6 minutes 34 seconds) that boys of the same age (9 minutes 41 seconds), teenage girls shower for a significantly longer time (9 minutes 21 seconds). On weekdays men spend 8 minutes 5 seconds showering, and although women shower in 7 minutes 39 seconds on average during the week, they spend higher time (8 minute 55 seconds) on Saturdays. According to the study, women do multi-tasking, like shaving legs and cleaning teeth, during showering (The Telegraph, 2011).

2.2.3 Attitudes, social norms, and behaviour

Based on the familiar concept of the 'Tragedy of the Commons' by Hardin (1968), a common pool of natural resources is overexploited in the absence of external control on individual behaviour. While the long-term effects of restraint can be beneficial for everyone, collective action is often not ensured due to individual rational choices guided by self-interest. Having conducted a series of experiments, Berk *et al.* (1980) suggested that the following conditions stimulate water conservation behaviour: (1) belief that a resource shortage exists and constitutes a problem for a group with which an individual identifies; (2) moral commitment to 'fair' contributions to group welfare; (3) belief in the efficacy of personal efforts to achieve a collective solution; (4) belief that the personal cost of inconvenience resulting from conservation efforts will not be great; and (5) belief that others in the relevant group will also conserve.

The theory of planned behaviour (Fishbein & Ajzen, 1975) has been used extensively to understand a range of behavioural decision-making contexts, including household water conservation (Kantola *et al.*, 1982; Harland *et al.*, 1999; Lam, 2006; Clark & Finley, 2007; Fielding *et al.*, 2010). The theory of planned behaviour is a parsimonious model of the informational and motivational influences that combine to predict behaviours. In this theory, the most immediate predictor of behaviour is an intention (i.e., a motivation or plan) to engage in the behaviour. The theory proposes intention to be determined by the additive effects of attitude, subjective norm and perceived behavioural control. Attitudes refer to the overall positive or negative evaluation of performing the behaviour. Subjective norms are based on individuals' perception of whether important other people in their life would want them to perform the behaviour. Perceived behavioural control reflects the extent to which individuals perceive the behaviour to be within their control.

Russell and Fielding (2010) used the term 'water conservation behaviour' to define the actions that aim to conserve water and capture the behavioural aspects of water demand management. The authors further differentiated between two types of water conservation behaviours: efficiency behaviours and curtailment behaviours. Efficiency behaviours refer to one-off behaviours, such as installing water saving shower heads or rainwater tanks, which facilitate ongoing water savings. In contrast, curtailment behaviours refer to individuals' actions that conserve water, such as only washing full loads of clothes, taking shorter showers and turning off the tap while brushing teeth.

In relation to water conservation behaviour, a range of studies based on the theory of planned behaviour found that attitudes predicted intentions to engage in water conservation curtailment and efficiency behaviours. Clark and Finley (2007) investigated curtailment intentions (e.g., taking shorter showers, sweeping terraces instead of washing them with water) and efficiency intentions (e.g., replacing existing shower heads and toilets with fixtures designed to use less water). In addition to the finding that people who had more positive attitudes to these actions had stronger intentions to engage in them, they also found that subjective norms (i.e., perceptions of social pressure to engage in water conservation), and perceived behavioural control (i.e., perceptions of control over conserving water) were also statistically significant predictors of water conservation intentions. That is, people were more likely to intend to engage in water conservation actions and install water efficient appliances if they thought that other important people in their life supported these actions and they felt that they were things that they could easily do. Similarly, Harland *et al.* (1999) showed that stronger intentions to turn off the tap when cleaning teeth were associated with more positive attitudes toward this behaviour, a sense that it was an easy thing to do, and personal norms that reflected a sense of moral obligation to engage in the behaviour.

Research by Lam (1999) showed that attitudes and perceived behavioural control were the most important drivers of efficiency intentions, and curtailment intentions. In more recent work in Taiwan, Lam (2006) attempted to predict people's intention to save water through the installation of dual-flush toilets. Lam found that the theory of planned behaviour alone is insufficient to understand people's intention to install dual-flush controllers in toilets, but that subjective effectiveness of alternative solutions is a good predictor for the intention. If the respondents believed that a drought would occur (perceived vulnerability) or that the retrofit could save water (response efficacy), they had greater intentions to retrofit. Lam's results suggest that water conservation campaigns should publicise not only the effectiveness but also the advantages of efficient technologies over traditional means. Similarly, Kantola *et al.* (1982) and Trumbo and O'Keefe (2001) found that attitudes and subjective norms were statistically significant positive predictors of water conservation intentions.

Fielding *et al.* (2010) used the theory of planned behaviour to identify the salient and important beliefs associated with household water use and conservation in South East Queensland. Three focus group discussions were conducted in each of four local government areas of Brisbane, Ipswich, Sunshine Coast and the Gold Coast. A semi-structured format was used to ask about the advantages and disadvantages, barriers and facilitators of, and important influences on, everyday water conserving behaviour and one-off installation of water efficient appliances. The participants of the study identified monetary savings associated with everyday conservation behaviours or installation of water efficient appliances as a major benefit of reducing water use. Most importantly, participants believed that installation of water efficient appliances has the advantage of lowering the need to monitor and be vigilant about everyday water using practices. According to the study, the major barriers of everyday water conservation included up-front costs and hassles of installing water efficient fixtures, concerns about functionality of these devices and the inconvenience arising from practicing water saving behaviour. The study concluded that the installation of water efficient appliances can be facilitated by providing free water audits and personalised advice and through rebates and assistance in purchasing and retrofitting existing appliances.

2.2.4 Water price

The water and wastewater tariffs implemented in cities across the world show significant variations in terms of monetary value as well as the pricing structure. Hoque and Wichelns (2013) studied the water and wastewater tariff structures in 60 cities across six regions and found that, on average, a household consuming 20 m^3 of water per month has a monthly bill of USD 42, which account for about 1.5% of the household monthly income. The price elasticity of demand for water is the most widely used indicator for understanding consumers' responsiveness to water price changes in a given context. Household demand for water is typically said to inelastic, meaning that an increase in price with result in a less than proportionate decrease in demand, leading to a steep demand curve. According to economic theory, demand for water tends to be price inelastic because: (1) it has no close substitutes; (2) the household expenditure on water bill comprise a relatively small share of the monthly budget; and (3) the demand for water is related to other complementary goods (Renwick & Archibald, 1998).

According to Beecher *et al.* (1994), the most likely price elasticity range for residential demand is −0.2–−0.4[1]. However, the price elasticity varies among regions, customer classes, water use categories, seasons, and time periods. Usually, price elasticity of demand for water is higher (more elastic) in the long-term as people have more time and resources to adapt their consumption patterns. It is also higher for non-discretionary outdoor uses such as watering gardens, filling up swimming pools and cleaning pavements and lower for more essential indoor uses. The responsiveness of demand to price change also depends on the tariff structure and the current rate level itself. If the tariff is progressive (i.e., unit rate increases with consumption) or if the original rate is very high, further increase in price will tend to have a greater impact on demand. Chapter 3 provides a detailed overview of the various water and wastewater tariff structures implemented worldwide and analyses their efficacy in demand management.

2.2.5 Dwelling type and ownership

The kind of homes people live in and whether they own or rent it not only influences overall water consumption levels, but also how people think about water use (Randolph & Troy, 2008). As illustrated by the Singapore example in Figure 2.2, households in landed or terraced properties use significantly more water than those living in apartments. This is because larger houses have more water use appliances, require more cleaning and may have significant outdoor water use activities. Researchers have shown that households who live in detached houses report greater intentions to conserve (De Oliver, 1999; Gilg & Barr, 2006; Clark & Finley, 2007). This is because a single family dwelling is in full control of its water consumption and any benefits accruing from conservation efforts will accrue to the household itself. In case of a multi-family dwelling unit, paying average water bills based on bulk meters, any reduction in total bill will be enjoyed by all households resulting in free-rider effect. However, in a study of residents in Taipei and Kaohsiung, Taiwan, Lam (2006) found that residents in detached houses had less intention to retrofit because they did not have to share water tanks with their neighbours, as did apartment residents. These findings suggest the need to be sensitive to cultural differences in how people use and conserve water.

Home ownership is another factor likely to influence water conservation behaviour. Randolph and Troy (2008) argued that home owners are likely to have direct control over their homes and are in a better position to undertake retrofitting through the installation of efficiency devices. In contrast, residential tenants have less control over the installation of water efficient devices and also do not necessarily receive a water bill as it is often a hidden cost as part of rental payments. In Australia, for example, only households in owner occupied houses receive water bills and pay for their water usage directly. Landlords can pass on water usage charges to the households of tenanted houses (provided the property has its own meter). The residents of housing units served by common meters, however, do not pay water usage charges directly. Instead, the strata corporation pays the charges and the costs are ultimately recovered through either strata fees or rents.

Existing research does not explore whether the influence of psychosocial variables such as attitudes and beliefs differ as a function of household tenure, nor does it investigate whether dynamics within households (e.g., between parents and children) impacts on residential water conservation. Indeed, Randolph and Troy (2008) and Grønhøj (2006) argued that household dynamics may play an important role in conservation behaviours through social norms and family dynamics, however, these propositions are yet to be tested empirically.

2.2.6 Regulatory policies

In environmental economics, there has been a long polarisation of views regarding the role of prices (market allocation) and the role of regulatory policies (government allocation) in controlling consumption and degradation of common pool natural resources. Traditionally, environmentalists view prices as ineffective due to the presence of market failures, the inability of monetary valuation to capture the true cost of natural resources and because of the relative inelastic demand for such goods and services. On the other hand, economists have full faith in price based approaches in controlling resource use due to lower transaction costs and the ability of prices to account for individual's willingness to pay for the environmental resource (Barrett, 2004). Chapter 6 of this book discusses the effectiveness of using price and non-price approaches towards promoting domestic water conservation.

Regulatory policies can be voluntary or mandatory and aim to affect individual behaviour either by command and control or by providing incentives. These regulations can be in the form of mandatory restrictions involving certain water use activities during specific times of the day or week or in the form voluntary conservation by providing consumers with informed choices. Many of these regulations or conservation programs have been largely successful in reducing water demand during periods of drought or in achieving long-term decline in overall domestic water use (see Chapter 4).

For example, Southeast Queensland (SEQ), along with other regions in Australia, suffered from the worst drought in recorded history during 2000s. The combined water level of the 12 dams (Grid Twelve), which form the primary water supply source for SEQ, fell below 40% in 2005 and continued to decline, reaching a minimum of 14% in mid-2007 (Figure 2.5). The Queensland Government's drought response comprised of a range of innovative infrastructure projects (such as the SEQ water grid), extensive institutional reforms, price increases as well as implementation of mandatory urban water restrictions (Spiller, 2008). The severity of these water restrictions increased as drought conditions worsened; Level 1 restrictions were implemented in May 2005 and were increased step by step to Level 6 in November 2007. As dam levels recovered since 2009, these restrictions were phased out and replaced with permanent conservation measures that should be followed at all times (Bureau of Meteorology, 2011). These restrictions, in conjunction with other price and non-price tools, significantly reduced domestic water consumption from 282 lpcd in 2004–05 to 143 lpcd in 2008–09 (Queensland Water Commission, 2010). As the restrictions eased, consumption gradually increased to about 180 lpcd in December 2012, still below the target of 200 lpcd (Vogler, 2013).

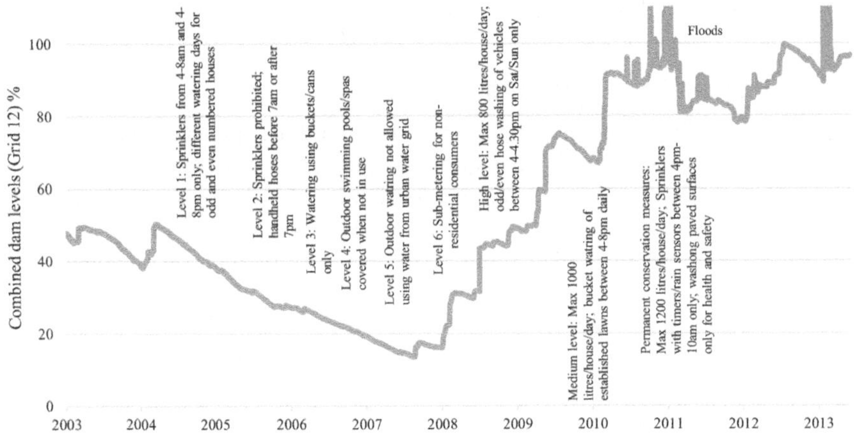

Figure 2.5 Dam levels and water restrictions in Southeast Queensland (Data on dam levels obtained through personal communication with SEQwater).

2.2.7 Technology

Reducing the water use of plumbing products – toilets, urinals, faucets, and showerheads – has long been a popular and effective conservation measure. However, initially some low-flow plumbing products met with controversy, primarily over their performance. When the US federal standards were set in 1992 (see Box 4.2), manufacturers released essentially the same products they always had, just adjusting the water level in toilet tanks or restricting the flow on showerheads. These products did not function well and resulted in many customers having bad experiences with 'low-flow' products. Since the 1990s, many manufacturers have made great strides in redesigning their products to work effectively at lower flush volumes or lower flow rates. In the beginning, the prices of these low flow fixtures have been quite high compared to their conventional counterparts; however, as technology advanced and the supply of these products increased, the prices have decreased significantly. The varieties of low flow fixtures available and their water conservation potentials have been discussed elaborately in Chapter 5.

In many cities, governments and water agencies try to promote the adoption of these products through public awareness campaigns, water efficiency labelling schemes and regulatory policies (see Chapter 4). When there is no significant price difference between a water-efficient and a standard appliance, it is often more expensive to retrofit the water appliances in an existing home than installing the same devices in a new house. The decision to install water efficient technologies in both existing and new properties are usually governed by factors such as household income, water tariff, attitudes and beliefs, awareness and ownership status of the dwelling (see Chapter 5).

2.2.8 Climatic variations

Changes in temperatures or seasons seem to have little effect on indoor water uses, especially for flushing, faucet use, cooking and cleaning. There may be slight differences in the amount of water used for showering as people tend to take more frequent and longer showers on hot days. In places where outdoor uses such as lawn watering contribute to a significant portion of domestic water use, the household water demand seems to increase during summer. For example, a 2010 annual survey of wholesale customers in Seattle found that while the average annual water consumption of a single family household is 145 gallons (548 litres) per day, the peak summer usage is 188 gallons (710 litres) per day (Seattle Public Utilities, 2011).

2.3 CONCLUSION

In this chapter, the main determinants of household water consumption have been discussed, although there are several other factors that can influence water usage at home. As mentioned in Section 1.6, cultural factors dictate the food preparation styles and clothes/dish washing methods (by machine or by hand). Water usage can also depend on the person performing the water use activities. In urban households, domestic helpers are usually engaged in all household chores and their water use behaviour may be very different from the residents who have different cultural, educational and socio-economic background. In Singapore, where a large number of domestic helpers come from Philippines, the Public Utilities Board (PUB) arranges training on water conservation at home through the employment agencies (PUB, 2013c). Water usage also depend on the amount of time spent at home, which, in turn, depends on the number of working people and school going children in the household. For example, in Singapore, general people usually have the practice of eating some meals outside, especially at hawker centres. Hence, a continued decline is household water consumption may be substituted by a corresponding increase in water use by the commercial sector. However, this issue has not been empirically proven yet.

Generally, outdoor water consumption usually refers to lawn watering or gardening, and excludes car washing. If cars are washed at gas filling stations, instead of washing manually at home, household water consumption with also decline. In my research regarding the potential of residential water conservation in Dhaka (Hoque *et al.* 2014), I found that upper middle class households usually have one car on average, which is washed 2 to 3 times a week, using hose pipes or bucket water. In all cases, it takes around 80–120 litres of water per wash. According to research by the International Carwash Association (Brown, 2012) commercial car washers use significantly less water compared to washing at home. Using a standard garden hose at a pressure of 50 psi for 5–6 minutes, requires about 50–60 gallons (189–227 litres) of water per vehicle. Self-service car washes (without reclaim system) use about 15 gallons, in-bay automatics use 30 gallons and frictionless conveyor washes use upto 70 gallons per vehicle.

Among the several factors discussed, not all factors have significant implication on household water use. While socio-economic and demographic factors are important, water pricing, regulatory policies and water-saving devices play the most significant roles in determining domestic water consumption and conservation patterns. Thus, this book focuses on the roles of prices, policies and technologies in greater detail, in the following chapters.

ENDNOTE

[1]Demand is said to be price inelastic when the value ranges between 0 and 1 and elastic when it is greater than 1. The negative sign implies the inverse relationship between price and quantity demanded.

Chapter 3

Role of prices: Water tariffs

Water as a natural resource may be available free of cost; but the extraction, treatment, storage and distribution of piped water entail huge amount of capital investments and operation and maintenance costs. These expenditures vary substantially from utility to utility depending on the proximity and quality of the water source, the infrastructure and technologies required to abstract, treat and distribute the water, the geographical expanse of the service area and the costs of raw materials and labour. For instance, cities that can source freshwater from nearby rivers incur much less costs than those that have to depend on water transported from outside the basin or country. Dependence on alternative sources involving desalination of seawater or recycling of wastewater may raise costs further. These differences in costs translate into differences in the water price that consumers pay for receiving water supply services.

This chapter discusses the role of water tariffs in promoting domestic water conservation. However, to understand the use of price as a demand management tool, it is necessary to recognise the role of tariffs in general and appreciate the wide variation in existing tariff structures. Section 3.1 provides a snapshot of the different components and rates of water and wastewater tariffs in selected cities across the world along with a comparison of household water and wastewater bills in these cities. Section 3.2 discusses the major objectives of the various types of tariffs and provides examples of cities where tariffs have succeeded or failed to meet local challenges. A detailed analysis of each of the tariff structures is provided in Section 3.3, particularly with reference to attaining some of the policy objectives discussed in the preceding section. Finally, Section 3.4 recommends how water pricing can be used as an effective tool for conservation, with specific emphasis on price elasticity of demand for water and representing information on water bills.[1]

3.1 OVERVIEW OF WATER AND WASTEWATER TARIFFS

The water tariff structures implemented in cities across the world show wide variations not only in terms of the nominal price per unit volume but also in terms of the different components into which the tariff is categorised. There can be different levels of categorisation in an urban water tariff schedule. Firstly, most cities have a separate tariff schedule for metered and unmetered connections, where metered consumers pay according to the volume of water consumed and unmetered consumers pay a flat rate per month. Secondly, tariffs are usually differentiated between broad consumer groups such as domestic and non-domestic. Thirdly, domestic consumers can be again subcategorised by dwelling types and non-domestic consumers can be separated by type of activity, such as government, commercial, or industrial.

The basic components that are usually included in a metered water and wastewater tariff schedule are described in Table 3.1. Some of these components have been analysed in detail in Section 3.3 of this chapter. These components can be combined in many different ways to design various forms of tariff structures.

Table 3.1 Components of water and wastewater tariff schedule.

Tariff component	Description
Basic service charge	
Fixed charge	This is a fixed amount paid per month or year and does not allow any minimum amount of consumption. The charge usually depends on the meter size and accounts for the cost of infrastructure and account maintenance. There can be combined or separate basic charges for water and wastewater respectively.
Minimum charge	This is a fixed amount paid per month and allows a minimum amount of free consumption of water.
Volumetric water charge	
Increasing block tariff (IBT)	This is a charge per unit volume which increases step-wise according to the level of consumption.
Decreasing block tariff (DBT)	This is a charge per unit volume which decreases step-wise according to the level of consumption.
Constant unit charges (CUC)	This is a charge per unit volume which remains uniform for all levels of consumption.

(*Continued*)

Table 3.1 Components of water and wastewater tariff schedule (*Continued*).

Tariff component	Description
Seasonal charge	This is a charge per unit volume which changes with the time of the year to account for peak (summer) and off-peak (winter) demands. It can be any of the three types listed above (IBT, DBT or CUC).
Wastewater or sewerage charge	
Volumetric charge	It can be any of the three types listed above (IBT, DBT or CUC) and the volume of wastewater generated is same as the volume of water consumed.
Flat	It is a fixed percentage (usually less that 100%) of the water bill. Mathematically, this can also be interpreted as a volumetric charge, but in this case a rate per unit volume is not specified.
Additional components	
Conservation or pollution tax	These are additional components implemented by a small number of utilities to account for the scarcity value of water or the environmental externalities caused by discharge of wastewater. These are usually a fixed portion of the total water bill.
Stormwater or property drainage charge	This is a fixed charge per month or year and depends on the land area of the property. It accounts for the fact that rainwater falling on a paved property ultimately discharges into the public sewers and increases the volume of wastewater to be treated.
Water resource development fee or capital contribution	Sometimes utilities impose a temporary fixed charge on consumers to earn revenues for development of additional infrastructure to meet expanding demands.

My colleague and I (Hoque & Wichelns, 2013) have conducted a study on urban water and wastewater tariffs in 60 cities in 43 developed and developing countries across six regions. As shown in Figure 3.1, half of the cities have a basic service charge for water supply and one-third have a basic charge for wastewater services. In case of volumetric water tariff, about 60% of the cities implement an

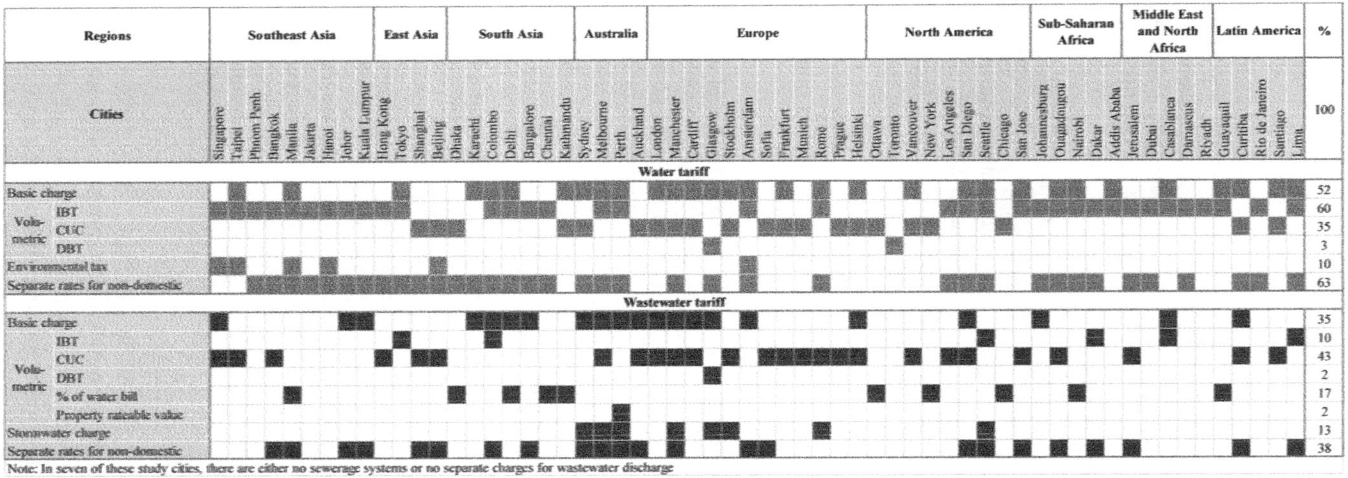

Figure 3.1 Components of water and wastewater tariff structures implemented in 60 selected cities (Hoque & Wichelns, 2013).

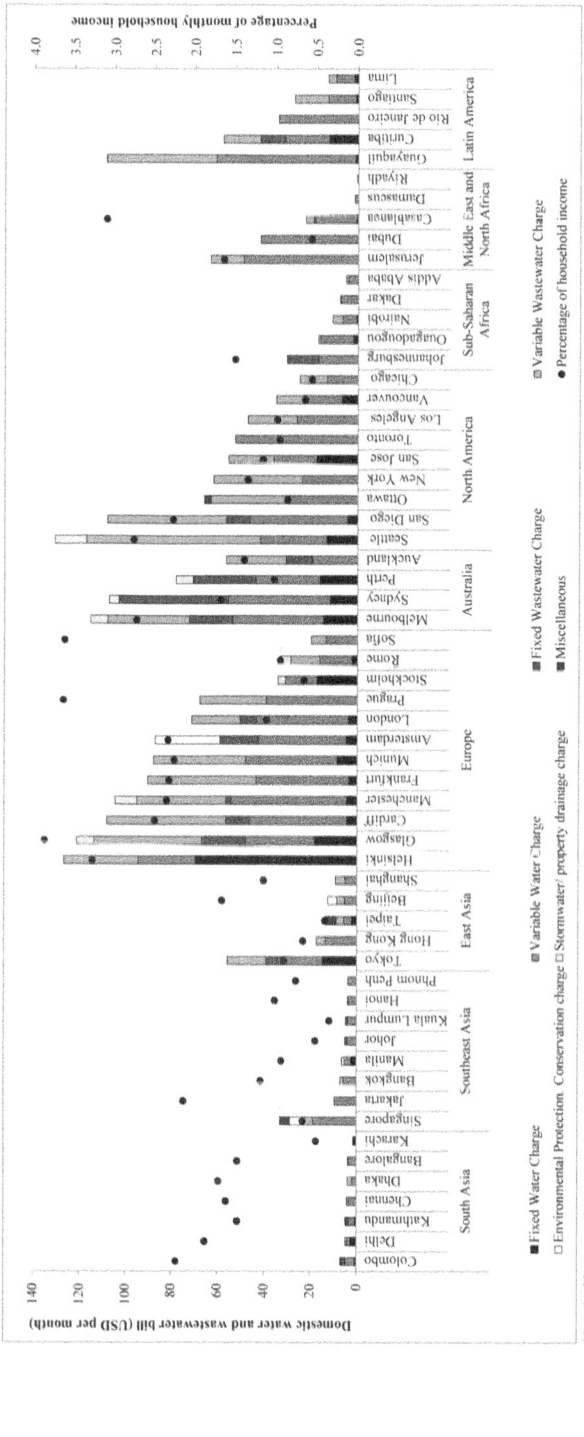

Note: In this paper, all currency conversions reflect 2011 average exchange rates (World Bank, 2011). Household income data is not available for nine cities in Latin America, Sub-Saharan Africa and MENA region.

Figure 3.2 Domestic water and wastewater bill based on a consumption of 20 m^3 per month (USD 2011 average exchange rates) compared with monthly household income (Hoque & Wichelns, 2013).

increasing block tariff (IBT), 35% use a constant unit charge (CUC) and only 5% has a decreasing block tariff (DBT). However, for volumetric wastewater charges, about 43% of the cities implement a CUC and 17% charge a fixed percentage of the water bill. Only 10% of the cities apply an IBT for volumetric wastewater tariff. A comparatively smaller number of cities implement a conservation or environmental tax, while stormwater drainage charges are only applied by some utilities in Europe and Australia.

As discussed in Chapter 2, the monthly household water consumption is primarily determined by factors such as number of household members, dwelling type or floor area of the house, climatic conditions, price of water, households' income, technological specifications of the water fixtures used and behaviours of consumers. Hence, monthly domestic water consumptions vary substantially between cities and also between different dwelling types within the same city, making it difficult to compare the average water and wastewater bills across cities. We have estimated the water and wastewater bill for a household consuming an average of 20 m^3 of water per month and expressed the bill as a percentage of the household income in each of the 60 cities (Figure 3.2). These are monthly bills for metered connections only and does not account for any discounts or benefits that consumers might be receiving.

Given the observed variation in tariff components and water prices across cities, generalised conclusions at the global or regional level could be misleading. However, a snapshot of the domestic water and wastewater bills from our sample reveals that the average unit bills in Asia and Africa generally are less than USD 1.00/m^3, and are lower than the average for all of the cities we consider (USD 2.10/m^3). In Europe, North America, and Australia, the unit bills generally are higher than USD 3.50/m^3. The GWI (2012) survey of domestic water and wastewater tariffs in 310 cities has shown similar results. Of the 51 cities with a unit bill greater than USD 4.00/m^3, 39 cities are in Western Europe, six are in North America, five are in Australia, one is in Latin America (GWI, 2012). Similarly, of the 81 cities with a unit bill less than USD 0.50/m^3, 49 cities are in south, east, and southeast Asia, while 13 cities are in the Middle East and North Africa, or in sub-Saharan Africa (GWI, 2012).

On average, households spend about 1.5% of monthly incomes on their water and wastewater bills. For some European, Australian, and American cities, the proportions are greater than 2%. In many of the Asian cities, although the monthly bills are quite small, the bills represent notable portions of the low levels of income earned in those cities. We use the average household income for each city when performing these calculations. Yet incomes vary substantially, particularly in developing cities.

3.2 ANALYSIS OF WATER AND WASTEWATER TARIFFS

Different forms of water tariffs can be effective in achieving different policy goals (see Section 3.3) and it is necessary to align the tariff components and rates with the local challenges faced by the water purveyors. This section discusses the advantages and disadvantages of the various water tariff components in relation to their general efficacy in addressing policy outcomes, along with examples of cities where each of the tariff components are being implemented.

3.2.1 Two part tariff: fixed + volumetric charges

This is one of the most common forms of pricing water, where a basic service charge is applied along with a volumetric charge. This variable charge can be an increasing block rate, a uniform rate or a decreasing block rate. The rationale behind the basic charge is that all connections impart a cost on the utility, due to installation of permanent infrastructure, such as pipes or meters, and related administrative costs. Fixed charges can account for these costs even when there is no consumption and reduce revenue volatility for utilities, while the volumetric charge gives the consumer some degree of flexibility in controlling their water bills based on their consumption. However, if fixed charges constitute a large proportion of the water bill, consumers have limited ability to control their bills and hence, have less incentive for conservation. Hence, the balance between the proportion of fixed and variable charges should be carefully determined based on the local priorities.

Some utilities do not have a separate fixed and variable charge as such, but they have a minimum charge that consumers need to pay for a certain quota of water consumed. Any consumption beyond the minimum quota is charged using volumetric rates. For example, in Kathmandu, domestic connections with a meter size of ½" pay a minimum charge of NRs. 55/month for consumptions of 10 m^3 or below and a uniform volumetric charge of NRs. 17.5/m^3 for consumption exceeding the minimum allowance (Kathmandu Upatyaka Khanepani Limited, 2009). Similarly, Manila Water charges a basic charge of PHP 101.01 per connection and allows free usage of 10 m^3 of water. Consumptions exceeding 10 m^3 is charged using an IBT structure ranging from PHP 0.28/m^3 for 11–20 m^3 to PHP 0.99/m^3 for consumptions above 150 m^3 per month (Manila Water, 2012).

According to the Independent Pricing and Regulatory Tribunal (IPART) of New South Wales, a two-part tariff schedule with a fixed charge and uniform variable charge is the most efficient pricing structure for monopoly service providers, as the variable charge can then be based on the marginal cost of supply and the fixed cost can recover the difference between the average cost and marginal cost (Cox, 2010)[2].

3.2.2 Single part tariff: increasing block tariffs

Increasing block-rate tariffs (IBT) are being adopted increasingly by utilities worldwide. With an IBT, the amount of water required for essential uses can be provided at a low price per unit, while consumers using large volumes of water per month are charged higher per-unit prices. The initial consumption block can be offered at a zero price, provided that the prices pertaining to higher consumption blocks are sufficient to generate the necessary revenue. In Johannesburg and Hong Kong, the first 6 m^3 of water consumed each month are provided at no charge.

IBT is usually referred to as a conservation oriented rate design, which also promotes revenue generation and cross-subsidisation between high and low-income consumers. However, to achieve the dual goals of water conservation and revenue generation, the sizes of the blocks and volumetric rates should be selected in accordance with local water consumption patterns. Wichelns (2013) shows that a three-part objective including affordability, revenue generation, and conservation can be achieved using an IBT if: (1) consumption in the initial pricing block is subsidised, (2) the volumetric rate in the second pricing block is sufficient to cover the operational costs and subsidies provided to consumers in the first pricing block, and (3) the volumetric rate in the third block is sufficient to cover both operational and investment costs.

One of the common arguments against using an IBT is that it poses a disproportionate burden on households with many members or where several households share a common connection (Whittington, 1992). In its early years, when Singapore decided to shift from a CUC to an IBT structure to account for the increasing costs of developing new infrastructure, it designed concessional rates for households with many persons. The general domestic tariff had four tiers, in which the first block of 0 to 25 m^3 was charged at SGD 0.22/m^3 (USD 0.17/m^3), the second block of 25 to 50 m^3 was charged at SGD 0.26/m^3 (USD 0.21/m^3), the third block of 50 to 75 m^3 was charged at SGD 0.33/m^3 (USD 0.26/m^3), and the remainder was charged at SGD 0.44/m^3 (USD 0.35/m^3). Concessional rates were designed for two or more households living in one block and for connections with 10 to 20 persons. While the rate in each block remained the same, the size of the blocks was increased for these consumers. The first block ranged from 0 to 50 m^3, the second from 50 to 100 m^3, the third from 100 to 150 m^3, and the fourth block was for consumption above 150 m^3 (Tortajada *et al.*, 2013).

An IBT may not be suitable for cities in which bulk meters serve high-rise buildings. One of the ways in which this issue can be solved is by using the average consumption per apartment to determine the applicable tariff rate. This approach has been adopted in Johannesburg, where the domestic water tariff consists of seven tiers. If a multi-dwelling residential connection comprising 10 households uses 100 m^3 per month, then based on the normal IBT rates, the total water charge would be $[(6*0) + (4*4.93) + (5*7.31) + (5*9.59) + (10*11.98) + (10*12.36) + (60*14.94)]$ = ZAR 1244.00 (USD 171.00) and each household will pay ZAR 124.40 (USD 17.10). However, based on the average consumption approach, the total consumption of 100 m^3 would be divided by 10 households, thus obtaining a consumption of 10 m^3 per household. The tariff for each household would be $[(6*0) + (4*4.93)]$ = ZAR 19.72 (USD 2.72) and the total bill would be just ZAR 197.20 (USD 27.20) (Johannesburg Water, 2010).

In theory, increasing block-rate tariffs can be designed to subsidise poor consumers who use small amounts of water per month. However, this goal is not achieved in some cities in developing countries, where many poor households in informal settlements are not connected to the water delivery system. Moreover, increasing the number of connections in poorer communities with lower water consumption will result in lower revenues for utilities, in comparison to increasing the number of connections to high volume users. This notion may serve as a disincentive for the utility to increase access to the poor, especially in the case of profit seeking private operators. This is exemplified by the case of Jakarta discussed in Box 3.4.

Increasing block-rate tariffs also are considered to be less efficient than uniform charges. In an IBT structure, the first block is usually subsidised and is priced at a rate lower than the marginal cost of service, while the additional blocks are priced at rates higher than the marginal cost. Moreover, all consumers, irrespective of their ability to pay, enjoy the subsidised rates for some portion of their consumption (Cox, 2010; Hoehn, 2011).

3.2.3 Single part tariff: constant unit charges

This is the simplest form of volumetric tariff that enables utilities to perform a straightforward calculation of water bills and enables consumers to easily comprehend and control their bills in relation to consumption. Apparently, a uniform charge seems to be less effective in promoting water conservation compared to IBT, but can be equally efficient in generating revenues if the rate is fixed at an appropriate level. Hoehn (2011), however, suggests that if a uniform volumetric charge (without any fixed charge) aims to cover both fixed and variable costs, the rate might become too high, and consumers might attempt to forgo beneficial water uses and waste time, money and resources in inefficient water saving.

3.2.4 Single part tariff: decreasing block tariffs

This tariff structure is appropriate in cities where abundant raw water sources allow average costs to decline when water supply increases. For instance, large industrial customers often enable the utility to capture economies of scale in water resource development, transmission, and treatment. Moreover, industrial users typically take their supplies from the larger trunk mains, and thus do not require the expansion of neighbourhood distribution networks (Whittington, 2003).

Well-designed decreasing block tariffs allow utilities to recover costs. However, they penalise consumers with low level of consumption and provide a disincentive for reducing wastage of water. There is a trend to move out of these kind of tariffs, essentially because water conservation has found place in the political agenda of many governments and marginal costs of providing water are now relatively high in many countries, hence decreasing block tariff are not any more profitable for utilities. A review of tariff policies in OECD countries revealed a continued decline in the use of decreasing block tariffs for households, in favour of a two-part tariff structure with a basic charge and a volumetric charge, such as CUC and IBT and a limited application of DBT among industrial users, especially large one, in a few countries (OECD, 2009).

For instance, in Glasgow, domestic connections with a meter diameter upto 20 mm pay GBP 2.14/m^3 for the first 25 m^3 of water consumed in a year and GBP 0.77/m^3 for any consumption above 25 m^3. For wastewater, the volumetric charge is 2.76/m^3 for the first 23.75 m^3 in a year and GBP 1.31/m^3 for the rest. These tariffs are charged on top of a fixed annual water charge of GBP 136.42 and a wastewater charge of GBP 140.10 (Scottish Water, 2012).

3.2.5 Single part tariff: seasonal water pricing

This type of pricing arrangement is primarily designed to promote water conservation during peak demand periods and also to account for the additional costs associated with supplying water when availability is low. Seasonal pricing can also increase revenues during peak demand periods (if consumption remains same) or prevent shortfall of revenue (if consumption declines due to conservation). It can be implemented in cities where there is significant difference in summer and winter water consumption and where water shortage becomes a critical problem during summer. Moreover, regular increases and decreases in tariffs constantly remind consumers of need for conservation compared to constant conservation charges all year round. One of the drawbacks is that changes in tariff rates need to be communicated to consumers constantly which may increase transaction costs. Box 3.5 provides a detailed examination of the seasonal water pricing structure implemented in Los Angeles.

3.2.6 Additional tariff components: conservation taxes

Additional charges such as water conservation fees, pollution charges or resource development fees are sometimes imposed by water utilities, on top of the usual water charges, to account for the shortage of water resources and costs of developing alternative water supply sources. Singapore is one of the countries using an IBT along with a water conservation tax and Box 3.1 shows the changes in the water tariff structure in Singapore.

Another example is Taipei, where a water resource conservation and compensation fee, amounting to 10% of the water charge, is collected by Taipei Water Department. The collected sum is applied specially for the infrastructure for water resources conservation and environment/ecology conservation in the water quality and quantity protecting area, as well as for feedback in public welfare and compensation to restricted land for local residence. This fee is charged on top of the basic service charge, the volumetric water charge, the sewerage charge and the pumping facility and maintenance charge (Taipei Water Department, 2011).

BOX 3.1 EVOLUTION OF WATER TARIFF STRUCTURE IN SINGAPORE

In Singapore, the water tariff and related taxes have been revised more than ten times between 1965 and 2012. While the initial price reviews were mainly motivated by the need to recover costs, tariffs reviews since the 1990s have been more focused on water conservation. Singapore shifted to an IBT pricing structure as early as 1973 and the range of the consumption blocks and the unit price in each block changed several times over the years. In 1991, the water conservation tax was introduced for the first time to discourage excessive consumption and was charged at 5% of the water charge for consumptions above 20 m^3 a month. This tax was subsequently increased to 10% and 15% in 1992 and 1995 respectively to control the growth in water demand in the domestic sector (Tortajada *et al.*, 2013).

Between 1997 and 2000, a series of changes were made to the water tariff structure with the aim of pricing water based on economic efficiency and making the tariff structure simpler and similar for all types of users. The volumetric water price was pegged to the cost of desalinated water to account for the higher costs of alternative water supply sources and the conservation tax was charged from the very first drop of water delivered to signify the importance of water as a strategic resource for the country. The IBT structure was changed from three tiers to two, while the conservation tax was changed from a flat percentage to one that increased with the consumption blocks.

Figure 3.3 shows the gradual adjustments in the water charge and the conservation tax during 1997–2000, after which the tariffs have remained unchanged till date. It should be noted that apart from these two components, the domestic water tariff in Singapore also consists of a sanitary appliance fee of SGD 3 per toilet in the house and a waterborne fee of SGD 0.3/m^3. Together with water pricing, the Singapore government adopted other complementary measures for water demand management (see Box 3.3) and as a result of these efforts, the per capita domestic water consumption in Singapore has declined over the years from 165 litres per capita per day (lpcd) in 2002 to 153 lpcd in 2011 (Public Utilities Board, 2011b).

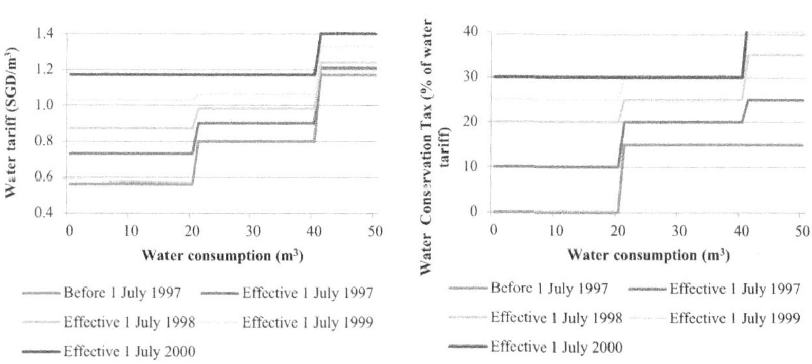

Figure 3.3 Changes in water tariff and water conservation tax in Singapore from 1997–2000 (USD 1 = SGD 1.26).

3.2.7 Income based measures: discounts or payment assistance

Given that water is a basic necessity and a human right, poor households cannot be deprived of access to clean water even if they cannot afford to pay the water bills. However, keeping the general level of tariffs low for all consumers can be counterproductive. Lower tariffs lead to decreased revenues, which, in turn, result in lower investment in water supply infrastructure, especially in poor peripheral areas of the city. Hence, instead of having a utility with ailing financial health, it is sometimes better to charge all customers at a cost effective rate and subsidise the low income households with targeted programs.

In Singapore, low income households, who have difficulty in paying their water bills, receive assistance in the form of Utilities Save (U-Save) rebates. This involves placing a fixed amount of cash in the household's utility account, which can be drawn at any time to pay any utility bill, including water; moreover, if the cash is not used up for one month, it stays in the account and can be used for subsequent months (Tan *et al.*, 2009). Thus, rather than having a low per unit charge for the first block under an IBT structure, which is enjoyed by all consumers irrespective of their income levels, this targeted subsidy achieves three purposes: assisting the poor, spreading the message that conservation is important even at low consumption levels and also earning revenues from all consumers who can afford to pay.

In London, customers that have been defined as 'Vulnerable' by the government, may be able to reduce their bill under the WaterSure scheme by only being required to pay the lower of their metered bill or GBP 333 (GBP 199 water and GBP 134 wastewater). Customers who may be eligible are those on certain specified state benefits who either have large families (there must be an entitlement to obtain child benefit for at least three children, in full time education, under the age of 19) or use significant additional volumes of water as a result of someone residing in the household having certain medical conditions (Thames Water, 2012).

3.2.8 Income based measures: subsidised network connections

Sometimes, even if the water tariff charged by the utility is much lower than alternative water supply sources such as informal vendors, poor households prefer not to connect to the formal network because of high one-time connection fees. On one hand, while excluding the poor from access to improved piped water supply poses a serious public health and equity concerns, on the other hand, it deprives the water utility from earning additional revenues by foregoing potential consumers. Moreover, this may also lead to a rise in illegal connections and increase non-revenue water (NRW) for the utility.

Subsidizing the connection fees can solve many of these issues, provided the utility has funds to do so. The case of Manila Water Company Inc. (MWCI), the private concessionaire operating in the east zone of Manila, shows how subsidizing network connections can prove to be a win-win situation for all. The company significantly reduced its water loss, from 63% in 1997 to 11.2% in 2011 and improved its reliability, in terms of 24-hour availability, from 26% of customers in 1997 to 99% in 2011 (Manila Water, 2011). The policies adopted by MWCI with regard to improving operational efficiencies, managing human resource, decentralizing business areas and increasing service coverage have been the main drivers behind this success. The company believed that connecting poor households to the network can significantly lower the NRW resulting from pilferage (Luz & Paladio-Melosantos, 2012). Hence, in its first year of operation, MWCI launched the 'Water for the Community', or *Tubig Para Sa Barangay* (TPSB) program, which was designed to rapidly increase connections to poor households by working through local community leaders to achieve communal solutions for water supply. In line with the program objectives, the land title requirement in applying for a water service connection was eased up, while flexible financing options and a socialised tariff scheme for low-income families were introduced. To date, over 700 TPSB projects have been completed serving 1.7 million people or 28% of the 6.1 million people served. In addition, the GPOBA is also providing subsidies to MWCI, through direct payment based on per connection installed in identified low-income areas. As surveys by MWCI revealed that the poor households can only afford to pay the meter and guarantee deposit of PHP 1620 (USD 40), the remaining connection fee of PHP 5911 (USD 144) will be paid by the subsidy, upon independent verification of satisfactory service delivery for 3 months (Global Partnership on Output-Based Aid, 2009).

3.2.9 Unmetered flat tariff structure

Volumetric water pricing through metering is one of the important pre-conditions to ensuring water conservation. When households are unmetered and charged a fixed assessed fee per month, there is no incentive to conserve water as any reduction in consumption will bring about no changes in the water bill. Yet, even in developed countries, water connections are largely unmetered, as exemplified by the case of the three cities in the United Kingdom in Box 3.2. Many of these household connections were established long ago and installing meters in these existing old houses may not be cost effective. This is particularly the case in cities where the benefits accrued from conserving water through universal metering may not be enough to offset the costs for installing, reading and maintaining meters. However, the percentage of metered connections is gradually increasing in many of these cities.

In Dhaka, in 2007, around 60% of the total connections was metered, while 40% was unmetered paying a fixed sum per month depending on the annual valuation of the property (DWASA, 2007). However, every year a number of connections are converted to metered connections and new buildings are always connected to a meter. Hence by December 2010, the percentage of metered connections rose to about 70%; of the total 4,379,350 connections, 3,037,094 were metered while the rest 1,336,605 were non-metered.

BOX 3.2 UNMETERED AND METERED DOMESTIC MONTHLY WATER AND SEWERAGE BILLS FOR 2012–2013 IN UNITED KINGDOM

The average monthly domestic water and wastewater bills are slightly higher for unmetered connections than for metered ones (Figure 3.4). The proportions of metered water connections are 33.4, 30.5 and 30.4% for Cardiff, London, and Manchester, respectively (OFWAT, 2011a). For unmetered households, the water charge includes a fixed annual standing charge and a rateable value charge, while the wastewater bill includes only the latter. For example, for water supply, unmetered households served by United Utilities in Manchester have a standing charge of GBP 59 per year (USD 93.65) and a charge of GBP 0.929 (USD 1.47) per GBP of rateable value of the property. For wastewater services, the charge is GBP 1.382 per GBP of rateable value of the property (United Utilities, 2012).

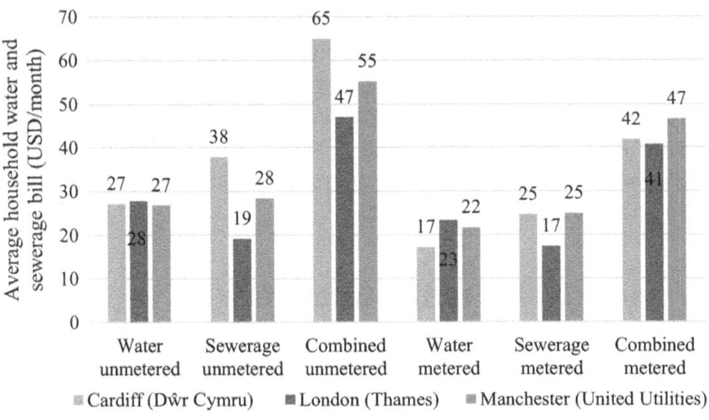

Figure 3.4 Monthly water and sewerage bills in three cities in the United Kingdom.

In Glasgow, although calculations based on a monthly consumption of 20 m³ yield a very high water and wastewater bill for domestic consumers, in practice, such high bills are rarely incurred as very few connections are metered. According to Scottish Water, the average annual household charge in 2012–13 was GBP 324 (USD 514). The majority of households in Scotland are levied Scottish Water's unmetered household tariffs; and as of March 2012, about 2,440,000 households were unmetered and only 497 were metered. While a metered domestic consumer in Band A (based on council tax on property) would pay GBP 904 (USD 1435) in 2012–13 for a consumption of 20 m³ per month, an unmetered consumer would pay GBP 262 (USD 416) for the same (Personal communication with Scottish Water, 2013).

3.3 OBJECTIVES OF WATER PRICING

The choice of components and their rates usually depend on the local challenges and the main priorities of the authorities. The best tariff design for a particular community and situation is one which strikes the most desirable balance among the objectives that are important to that community (Boland, 1997). Consumers and suppliers of water have different expectations of water tariffs. Consumers like high quality water at an affordable and stable price. Suppliers like to cover all costs and have a stable revenue base. The level and structure of fees for water and wastewater services have consequences far beyond these expectations. Water-related fees can be expected to generate revenue, improve efficiency of the supply and supplier, manage demand, facilitate economic development and improve public welfare and equity (Potter, 1994). Whittington (2003) highlights that the four major objectives that should be taken into consideration while setting tariffs are revenue sufficiency, economic efficiency, equity, and poverty alleviation. The following paragraphs discuss some of these objectives, with examples of cities where particular tariff designs have succeeded or failed to achieve the desired objectives.

3.3.1 Revenue sufficiency

The main aim of water prices is to generate sufficient revenue to cover the recurrent expenses and to provide funds for capital investments as well. However, unlike other services such as electricity, gas or telecommunications, water is a basic amenity of life and a human right, which makes setting water tariffs a highly sensitive political issue. Basic economics require that the price of a service be at least as high as the cost of providing that service. Rogers *et al.* (1998) argued that sustainable and efficient use of water require the tariff to match not only costs of supply (i.e., O&M and capital costs), but also opportunity costs, economic externality costs, and environmental externality costs. However, most often, the water tariff is just enough to meet the O&M costs and capital expenses are mainly subsidised by the government or sourced from external donors. Other intangible costs involving environmental externalities and opportunity costs of diverting water from alternative uses are largely ignored. Box 3.3 shows how water tariffs have been restructured in Tokyo to account for the changes in consumer types.

3.3.2 Affordability

While tariffs need to be high enough to cover costs of supply, public authorities are often reluctant to raise prices to cost effective levels based on the rationale that higher expenditures on water bills would affect the affordability of the poor and may lead to political unrest. For this reason, water tariffs are kept very low, especially in developing cities, where a large section of the population are low-income residents, often living in informal settlements. However, in most cases, the keeping the general level of tariffs low lead to counter-productive outcomes. Lower tariffs result in poor financial health of the utilities; lack of funds deters expansion or provision of services to low-income areas growing along the periphery of the city. As a result, poor residents are not connected to the piped water network and fail to receive the benefits of the lower tariff. These residents then have to supplement their water needs by purchasing water from informal vendors at a much higher cost. Box 3.4 discusses the tariff structure in Jakarta and analyses its effectiveness in ensuring affordability for the poor.

BOX 3.3 RESTRUCTURING TARIFFS TO MAINTAIN REVENUES IN TOKYO, JAPAN

In September 2004, the Tokyo Metropolitan Assembly approved of a new water tariff structure proposed by Tokyo Metropolitan Water (TMW). The current water tariff structure which has been enforced since January 2005, is 2.2% lower than the previous water tariff that had been effective since 1994. This was the first time to propose water tariff reduction in the history of TMW. The background of such reduction is the changes of socioeconomic and water use structure in Tokyo; bulk users, such as factories and small and medium-sized enterprises, have been gradually getting out of Tokyo or have just closed their businesses, and the average household size as calculated by the number of members per household has been gradually downscaling (Takizawa *et al.*, 2005). The progressive water tariff in Tokyo, the reduction of bulk users and increase of small users, as well as the decrease in total water supply volume, had a direct impact on the revenue of TMV. Thus, TMV reorganised its tariff structure to account for the socioeconomic changes. Compares the current tariff structure (2005–present) with the previous one (1994–2004) (Table 3.2).

Table 3.2 Fixed and volumetric water charges for 13–25 mm meters in Tokyo.

Year	Basic monthly charge (USD)	Volumetric Charge (USD/m³)								
		0–5	6–10	11–20	21–30	31–50	50–100	101–200	201–1000	>1001
1994	15.41	0.00	0.00	1.63	2.19	2.69	2.69	3.76	4.70	5.20
2005	14.66	0.00	0.28	1.60	2.04	2.53	2.67	3.73	4.66	5.06

Source: Takizawa *et al.* (2005); and Tokyo Waterworks Bureau (2005)

BOX 3.4 WATER PRICING IN JAKARTA, INDONESIA

Since 1997, the water supply services in Jakarta are managed by two private companies, PT PAM Lyonnaise Jaya (Palyja) for west zone and PT Aetra Air Jakarta (Aetra) for east zone, under two 25 year concession contracts, in which PAM Jaya, the previous public utility is the owner and represents Jakarta authority. Compared to other Asian capital cities, Jakarta has minimal water supply and wastewater disposal

(*Continued*)

BOX 3.4 (*Continued*)

infrastructure, with service coverage of 63.5% and 65.8% and NRW of 39% and 45.5% in the west and east zones respectively (Palyja, 2011; Aetra, 2013). Most of the network connections are spatially concentrated in the high income areas and even those connected to piped water often have to rely on other back-up sources due to low water quality and intermittent supply. Those who are not connected to the network depend on informal vendors, deep and shallow tube wells, and/or bottled water (Bakker & Kooy, 2012).

Lack of finance is the main impediment to expanding coverage and improving service, and the current pricing structure and financial payment mechanism between PAM Jaya and the private operators have worsened the situation. Jakarta has an IBT structure differentiated between seven customer categories (Figure 3.5), based on income levels, where the area of the house is taken as a proxy for income.

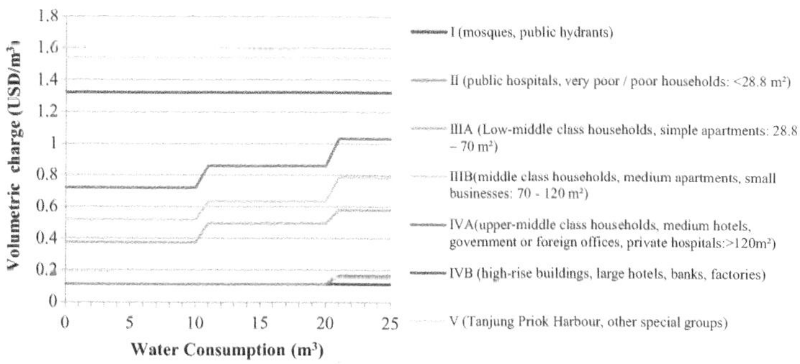

Figure 3.5 Water tariff structure in Jakarta (PAM Jaya, 2007).

Although the main rationale behind this structure has been to lower the burden of water bills on the poor and to allow the richer households to cross-subsidise for the low-income ones, in practice, this has not been fully achieved. This is because the proportion of high-income households and industrial consumers is very less compared to the low-income households and hence, the revenues collected from the former are not enough to compensate for the latter. Moreover, poor households are mostly not benefitted from this differentiated tariff structure, as many are not connected to the piped network in the first place.

According to a 2005 survey of 110 poor households in six neighbourhoods in Jakarta, vendor water was 10 to 32 times more expensive than piped water and 43% of surveyed households spent more than 5% of their income for water (Bakker & Kooy, 2012). Yet, the poor are usually reluctant to apply for formal network connections due to high one-time connection fees (USD 66 for Groups I and II, USD 101 for III, and USD 122 for IV) (PAM Jaya, 2007), lack of proper land tenure and additional coping costs to deal with intermittent supply. Moreover, as low-income households have fluctuating daily incomes, they have limited budgeting capacity, which favours the flexible credit based payment options offered by the vendors (Bakker & Kooy, 2012).

According to the concession contract, PAM Jaya is liable to pay the private operators a 'water charge' based on the total volume of water sold and billed and

(*Continued*)

BOX 3.4 (*Continued*)

it is adjusted every semester to account for inflation rates and foreign currency adjustment. However, soon after the contract was signed in 1997, the East Asian financial crisis caused massive devaluation of the Rupiah and severe political unrest, which made the Governor unwilling to implement agreed-upon tariff increases. This three-year tariff freeze caused the water charge to rise beyond the average tariff revenue received, thus, increasing the debts of PAM Jaya. Moreover, increase in operating costs arising from imported materials and decrease in payments received from PAM Jaya, prevented the private operators from realising their investment and technical targets. This led to down-scaling of contractual targets; and despite several negotiations and multiple tariff increases between 1998 and 2007, the debts could not be repaid (Jakarta Water Supply Regulatory Body, 2009).

Given this critical financial condition, both PAM Jaya and the private operators are reluctant to increase connections to poor areas, as it will only increase the debt burden further. Thus, in 2003, 87% of network connections were provided to Categories III and above, whereas proportionately more residents belonged to lower-middle or low-income categories (Bakker & Kooy, 2012). Hence, although de-linking the private operator from collecting tariffs directly from the consumers aimed to remove the disincentive to increase connections to low-income households, this payment mechanism through water charges resulted in counterproductive outcomes, due to increasing government debt.

Having recognised these barriers, the private companies have undertaken limited initiative to improve access to water supply for poor communities. One of these pro-poor initiatives involves placing master meters in illegal settlements, from which individual households can connect themselves through community based organisations. Palyja has also allowed consumers to pay their connection fees in 12 monthly instalments, as part of their first year water bills. Moreover, the Global Partnership Output based Aid (GPOBA), funded by the World Bank, has enabled Palyja to provide connections to about 5000 poor households in slum areas until 2012 (Palyja, 2012).

For the last five years, the water tariffs in Jakarta have not been increased, as according to the Governor, the private companies have not fulfilled their contractual targets. In their 2011 Annual Report, Palyja has clearly stated that this tariff freeze has heavily impacted their investments and service delivery, although their financial statements show large amounts of profits. Given this scenario, experts and policy advocates are pushing for terminating the contracts, as the primary objectives of private sector participation – greater access to finance to improve service coverage and quality – have only been partially materialised.

3.3.3 Equity

Apart from addressing the primary objectives of revenue sufficiency and affordability, water tariffs are also expected to ensure equity and fairness to all consumers. Equity means that the water tariff treats similar customers equally, and that customers in different situations are not treated the same. This would usually be interpreted as requiring users to pay monthly water bills that are proportionate to the costs they impose on the utility by their water use (Whittington, 2003). Earlier, it was thought that IBTs can address the issue of equity, by requiring poor households that consume less water to pay lower unit price than richer households that consume larger volumes. However, the problems with IBTs are now apparent – poorer households often share connections and have large families, leading to higher consumption at the household/ connection level (see Section 3.2.2 above).

It is difficult to formulate a universal criterion for equity as ensuring fair distribution entails philosophical and social considerations and value judgements. Barberán and Arbués (2009) notes that there are two approaches to equity: (1) the ability-to-pay principle, which justifies a preferential treatment of low-income household to meet their basic needs; and (2) the benefits principle, which requires payments to be proportional to the benefits received (which can be interpreted as the amount of water consumed). However, as benefits are accrued at an individual level, while bills are paid at household level, Barberán and Arbués (2009) propose that to ensure equity, all individuals should be able to meet their basic needs at the same per capita variable charge, regardless of the household size.

3.3.4 Demand management

Finally, an appropriately designed tariff structure can be an important tool for managing water demand, provided other aspects of governance are in place. The efficacy of using price as an instrument to influence water usage is a highly contentious issue, given the relative inelastic demand of water. Given that water is a necessity for everyday household activities and expenditures on water bills comprise a very little portion of the households' monthly income, consumers tend to be quite unresponsive to price signals. As discussed in Chapter 2, there are a large number of studies estimating the price elasticity of demand of water for domestic uses; and although they vary in terms of the methodologies, data-sets and the actual numerical values, the general consensus is that household water demand is inelastic. Box 3.5 shows pricing policy adopted in Los Angeles to promote water conservation during the summer.

BOX 3.5 SEASONAL WATER PRICING TO CURB PEAK DEMAND IN LOS ANGELES, UNITED STATES

In 1993, the Los Angeles Department of Water and Power (LADWP) restructured its water rates to provide customers with a clear financial signal to use water more efficiently. LADWP introduced a two tiered seasonal pricing structure, where higher volumetric rates are charged during the high season (summer) compared to the low season (winter). According to Hanemann (1993), one of the stakeholders involved in designing the new tariff structure, the two-tiered structure was aimed at shrinking the right-hand tail of the distribution of demand by having a substantial price differential between the two clocks and locating the switch point at a level of usage where demand might be reasonably responsive to price incentive. In order for the price incentive to work, Hanemann believed that it was not necessary to make everyone pay the

(Continued)

BOX 3.5 (*Continued*)

higher rate on some units of their consumption; rather, the incentive would still be effective for consumers below the switch point, as long as it was sufficiently close that the higher price loomed in their consciousness and could influence their purchase of water saving appliances. On the contrary, tariff structures with large number of blocks and quite small differentials between the blocks often dilute the incentive effects.

Unlike other utilities where the boundaries for the consumption blocks/tiers are fixed for all consumers and all year round, a unique feature of this rate structure is that the first tier allocation considers factors that influence individual residential customer's water use patterns. This type of arrangement not only sends an effective price signal for conservation, but also addresses equity, provides basic water needs at an affordable price and generates adequate revenue for maintaining and upgrading the water system. Moreover, similar to few other cities, the water bill in Los Angeles is solely based on water consumption, and does not include any fixed costs.

For single-family residential customers, the breakpoint between first and second tier usage is based on lot size (five categories), temperature zone (three zones), and household size (the household adjustment involves a sliding scale of allowed extra first tier usage for households of seven to households of thirteen or more) (Table 3.3). For multi-family residential customers, and commercial, industrial and governmental customers, the Tier 1 allotment is based on a percentage of the customer's maximum daily average during from the previous December through March.

Table 3.3 First Tier Usage Blocks (Hundred cubic feet per month) for single family residential consumers in Los Angeles.

Lot Size Group (ft^2)	Season	Temperature Zone		
		Low	Medium	High
			(100 ft^3 per month)	
1–7499	High	16	18	19
	Low	13	14	14
7500–10,999	High	23	26	27
	Low	16	17	17
11,000–17,499	High	36	40	42
	Low	24	25	25
17,500–43,599	High	45	51	53
	Low	28	29	29
43,560 and above	High	55	62	65
	Low	36	38	38

Notes: 1 ft^2 = 0.0929 m^2.
1 hundred cubic feet = 2.83 m^3.
High season: 1 June to 31 October.
Low season: 1 November to 31 May.
Temperature zone is determined by the location of the property (zip code).
Source: Los Angeles Department of Water and Power (2012b).

(*Continued*)

BOX 3.5 (*Continued*)

During periods of drought, shortage year rates are implemented during which the switch-point between the first and second tier is reduced by 10%, 15% or 25% to encourage people to practice additional water conservation and also to offset any revenue losses resulting from lower consumption. Since June 2009, shortage year rates became effective, under which first tier allotments are reduced by 15%. In 2012, the monthly amount of water allotted for single dwelling residential units of land area less than 7500 ft^2, living in a medium temperature zone is 1200 and 1550 ft^3 for low and high seasons respectively. Hence, based on an assumption of a household consumption of 20 m^3/month (706 ft^3/month), all consumers fill fall within the allotted amount and pay the tier 1 rates. The amounts in each tier are based on a 60 day billing cycle (Los Angeles Department of Water and Power, 2012a). Figure 3.6 shows the Tier 1 and Tier 2 rates of 2011–12.

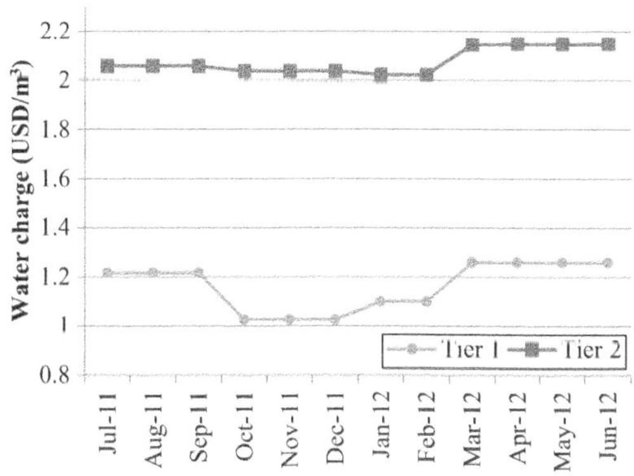

Figure 3.6 Seasonal water pricing structure in Los Angeles (Los Angeles Department of Water and Power, 2012a).

3.4 WATER PRICING AS A TOOL FOR CONSERVATION

3.4.1 Designing a conservation oriented rate structure

Conservation-oriented water pricing is a rate structure adopted by a water service provider where the costs of providing services are recovered, individual customers are metered and pay for the volume of water they use, and the price signal is sufficient to affect individual decisions and encourage conservation and efficiency (Brandez *et al.*, 2010). The most commonly used conservation oriented rate structures are the increasing block tariffs, seasonal rates and/or the inclusion of a water conservation tax. Moreover, there are some alternative forms of rate structures, which have strong theoretical basis, but are rarely used in practice. One example is the 'sliding scale rate' in which the price per unit of all water used increases based on average consumption. It is mainly aimed to reduce average and peak demand. Another possibility is 'scarcity pricing' in which the price of water per unit volume increases as water availability decreases in the short run and again decreases when supplies are replenished. This is aimed to price water using its short run marginal cost rather than the long-run marginal cost. Box 3.6 discusses the potential of using scarcity pricing in the water sector in Australia.

BOX 3.6 POTENTIAL OF SCARCITY PRICING IN AUSTRALIA

In Australia, regulators and water businesses have attempted to provide signals for efficient water use by setting the volumetric rate with reference to the long run marginal cost (LRMC) of supply. The LRMC can be defined as the cost attributable to an extra permanent unit of consumption in bringing forward the future capital program. LRMC prices signal the future costs of capacity augmentation to meet growth over the longer term but do not respond to increasing scarcity of water and do not reflect the high degree of variability in inflows in the short run.

During the millennium drought, that led to severe decline in water levels in storage dams in most parts of Australia, governments and water businesses turned to water restrictions and conservation programs in an attempt to rebalance supply and demand. As the drought continued, governments and water businesses invested billions of dollars in supply augmentations, particularly through high-cost, high reliability manufactured water sources, mainly desalination plants. While primary reliance was placed on demand management and investment in new supply rather than pricing to restore the supply–demand balance, there was increasing adoption of increasing block tariffs (IBTs).

Growing recognition of the significant economic and social costs of restrictions as a means of rationing limited supplies and the fluctuations in precipitation due to climate change have led governments and regulators to consider 'scarcity pricing' as a more efficient means of allocating water to uses most highly valued by users and the community. Scarcity pricing involves setting a price for water that would be higher when water was relatively scarce (e.g., dam levels were low) and lower when water was more plentiful (e.g., when dam levels were high). Scarcity pricing can be a more economically efficient way of managing short-term scarcity than water restrictions that do not adequately take into account the value individuals place on certain uses of water.

However, there are some common concerns with regard to the effectiveness of scarcity pricing. It could lead to excessive variability in prices, which would send confusing signals to customers and revenue volatility to water businesses. LRMC pricing is simpler, more transparent and stable. Scarcity pricing could sometimes result in very high prices, which would raise concerns about consumers' ability to pay. Restrictions are more equitable and supported by customers. Scarcity pricing may give an incentive for monopoly water businesses to artificially generate shortages in order to make excess profits. Scarcity pricing may be effective than restrictions in generating reductions in water consumption because demand for water is inelastic. Related to this is the absence of intelligent networks (i.e., smart meters), infrequency of billing cycles, the significant proportion of direct customers not receiving water bills (such as tenants), and the fact that water bills account for a small proportion of total household expenditure.

Source: National Water Commission (2011).

The design of a conservation-oriented rate structure for residential customers should take into account the responsiveness of water usage to water rates, that is, price elasticity of demand. According to Beecher *et al.* (1994), the most likely price elasticity range for residential demand is –0.2 to –0.4, which implies that an increase in water rates will lead to an increase in utility revenues. However, the price elasticity varies among regions, customer classes, water use categories, seasons, and time periods. It is recommended for utilities to estimate price elasticity using the data specific for their own customers.

Generally, water demand for outdoor discretionary uses (such as lawn watering, car washing, and swimming pools) is more elastic than the demand for non-discretionary indoor water uses. Water demand of low-income customer groups is usually less elastic than the demand of high-income customer groups because low-income customers usually use less water for non-discretionary uses (Beecher *et al.* 1994).

Price elasticity of demand also depends on the tariff structure and the rate level. If tariff rates are low in the first place, an increase in the rates from block to block would have a small effect on quantity demanded. The rates are low if the total water bill amounts to a very small share in the total household income. Water demand is usually more price-elastic in the long run (about five years or longer) than in the short run. In the long run, customers have the opportunity to adapt to higher prices by making their appliances and outdoor irrigation methods more water efficient. Stevens *et al.* (1992) have conducted a cross-sectional analysis of water demand for 85 communities in Massachusetts in 1988. The authors employed an average price variable to calculate the price elasticities of three commonly used volumetric rates structures. The results showed that the price elasticities ranged from –0.10 to –0.43 for uniform rates, from –0.40 to –0.69 for decreasing block rates and from –0.42 to –0.54 for increasing block rates. The study concluded that while price elasticities were not significantly affected by rate structures, the level of rates were more important in influencing water consumption. Similarly, Martin and Thomas (1986) used data from 1978–79 to conduct a cross-sectional analysis of residential water demand for four cities in Tucson and Phoenix, Arizona. The study found that the long-run price elasticity was of about –0.50 over a wide range of prices and indicated that residential water demand tends to become more price-elastic with higher water prices.

While improvements in a household's water appliance efficiency can lower its average water demands, it can also reduce its responsiveness to changes in water usage prices. A study on residential price elasticity of demand for water in Sydney, conducted by Saratidis (2011) in collaboration with Sydney Water, analysed a sample of around 95,000 individual households and 3300 blocks of housing units through time. The study found that once a household has upgraded the efficiency of its water use appliances (e.g., a four star washing machine) its long-term price elasticity is almost halved.

3.4.2 Communication of the price signal through consumer billing

According to microeconomic theory, the demand for water should decrease as price increases. However, this law of demand implicitly assumes that consumers know prices, an assumption that is not always satisfied in markets with *ex post* billing (Gaudin, 2006). When prices are not transparent, elasticity estimates are potentially lower than their full information potential. In order to study the effect of price information on residential water demand, Gaudin (2006) obtained information on water bills from 383 utilities across the United States of which 130 charged decreasing block rates, 104 increasing block rates, and 149 uniform per unit rates. The kind of information given on water bills varied from city to city along several dimensions. While all bills included meter readings, quantity used, and total amount due for water (separate from other charges), there was significant variation on whether the bill included marginal price information and history of use. Only 17% of utilities in the sample indicated price per unit next to consumption and an additional 3% indicated the price schedule somewhere on the bill. For history of use, 23% of utilities provided a simple comparison to the same period last year was the most common, while only 6% gave more extensive historical data (multiple months, generally presented with a graph). About 10% of utilities gave messages on conservation while another 10% supplies other information such as benchmark comparisons, daily consumption, and percentage changes in consumption. Based on results obtained from a regression analysis, the author concluded that price information significantly increases the elasticity of demand for water; all other factors held constant, a utility that gives marginal price information on the water bill can attain the same level of conservation with a 30–40% lower rate increase (Gaudin, 2006).

In Singapore, households receive a monthly utility bill from Singapore Power (SP Services). The bill gives a breakdown of the water bill into its different components, along with graphs showing the household's water consumption in the past six months in comparison to the national average for similar dwelling types (Figure 3.7). However, the invoice also contains electricity and natural gas bills, which make up the bulk of the total – ranging from SGD 35 (USD 27) in single-room apartments to SGD 695 (USD 551) in bungalows (Singapore Power, 2013b). Households are thus more concerned about electricity consumption, and might not be looking separately into their water bill. Many consumers will not even notice the water portion of their bill, as the charges are deducted directly from their bank accounts. Combined billing reduces administrative costs compared to sending three separate meter readers and bills; it also reduces transaction costs for households. Yet the practice is probably not effective in communicating information regarding water prices (Hoque & Wichelns, 2013).

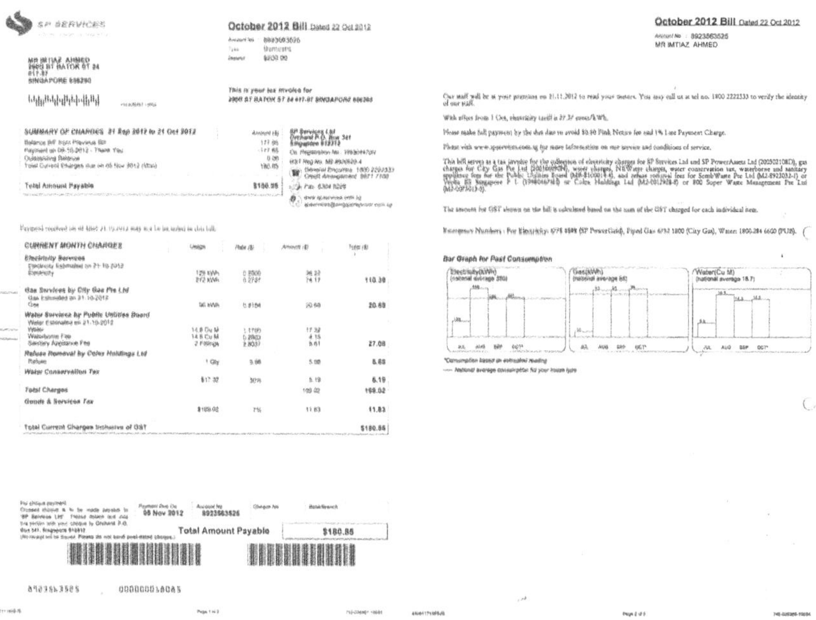
SP SERVICES

October 2012 Bill Dated 22 Oct 2012

October 2012 Bill Dated 22 Oct 2012

MR IMTIAZ AHMED

Total Amount Payable $180.85

Figure 3.7 Sample electricity, water and gas bill from SP Services, Singapore.

In the UK, while the percentage of metered connections is very low compared to other developed countries, the billing procedure for those that are metered is also not effective in informing consumers about their monthly consumptions. As mentioned in OFWAT (2011b), these 'dumb' meters, which are usually located in the basement of the dwelling or other places inaccessible to the consumers, have no facilities to store information and are read manually by water companies every six months or once a year. Hence, despite having a metered connection, households pay a fixed 'estimated' bill, based on the number of people in the house and dwelling type (for new occupiers) or on consumption during the previous billing cycle (for existing occupants). Following the next meter reading, money is either refunded or charged based on the difference between the actual consumption during the billing cycle and the estimated bill. 'Automatic meter readers (AMR)' coupled with data loggers, which are more expensive, can transmit information to water companies and thus avoid the need for manual reading and estimated bills. In New York, the Department of Environmental Protection (DEP) is automating its water meter reading capabilities to increase billing accuracy and provide customers with the tools they need to better manage their water use. DEP is also providing AMR-installed customers with an online application that enables property owners to view and manage their consumption on a daily, weekly, monthly, and yearly basis (NYC Environmental Protection, 2013). An even better and more expensive option involves 'smart meters', which have in-house displays and can transmit water usage data in real-time, allowing disaggregation of water consumption by appliance or activity. Currently, these meters are only used for research purposes.

Sarafidis (2011) in association with Sydney Water, mentioned in Section 3.4.1, estimated the immediate and long-term responsiveness of residential households to water usage prices for three categories of user groups: owner occupied houses, tenanted houses and housing units. These user groups were chosen because of the way they are billed for water and wastewater services. Households in owner occupied houses receive water bills and pay for their water usage directly. Landlords can pass on water usage charges to the households of tenanted houses (provided the property has its own meter). The residents of housing units served by common meters, however, do not pay water usage charges directly. Instead, the strata corporation pays the charges and the costs are ultimately recovered through either strata fees or rents. As the latter two types of households do not receive or pay their bills directly, they are often ignorant to price signals. This is substantiated by the price elasticity of demand of water, which is −0.08 for owner occupied houses, −0.02 for tenanted houses and −0.01 for housing units with common meters (Sarafidis & Sydney Water, 2011).

A much earlier study by Agthe *et al.* (1988) on a sample of 542 households in Tucson, Arizona found that 248 households did not know that Tucson Water had used an increasing block rate pricing structure for the preceding seven years, 75 were aware of the block rates and believed they reduced water use, while nine were aware but believed the rate structure had no effect. This is because the water bills provided only very limited encoded information while the block rate structure was not presented at all. The complexity of the rate structure also confused the customer and prevented acquisition of rate structure knowledge. The water rate structure included a monthly service change, several block rates, a summer surcharge, and a sewer use fee that includes a monthly service charge and an additional charge based on water volume from the prior winter months. According to the authors, a simplified tariff structure and greater information, in a more readily digested format, could be included on the water bill to help customers make rational economic decisions concerning water use.

One of the current examples of a complex tariff structure is that of Manila Water and Maynilad, the two private concessionaires operating in the east and west zones of Manila respectively. The tariff schedule of these companies include six components – the basic fixed charge per connection, the volumetric charge, the foreign currency differential adjustment, the environmental charge, the sewerage charge and the maintenance service charge. Of these, the volumetric charge for residential consumers consists of nine tiers with the difference between the highest and lowest tier rate being USD 0.70. Similarly, the business group 1 has 33 tiers with a difference of only USD 0.10 between the first and last tier. However, this complex tariff schedule and bill computation steps were completely missing from the water bill until September 2012, when the on-site bill printing system was launched. Although the latter also does not show the tariff schedule, it gives information about the charges under each component and the water consumptions of the three preceding months (Figure 3.8).

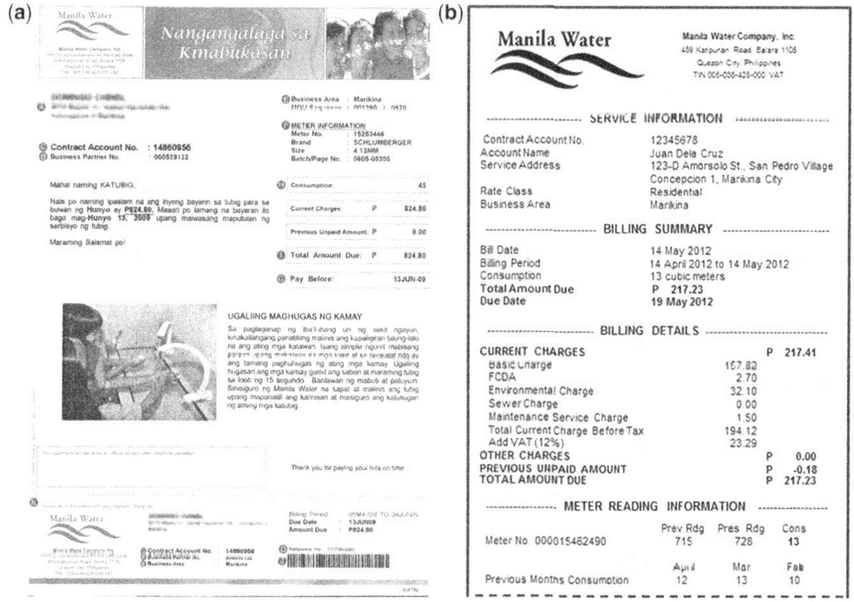

Figure 3.8 Sample water bill from Manila Water before (a) and after (b) September 2012.

3.5 CONCLUSION

Water conservation pricing is the use of rate structures that encourage consumers to use water efficiently, while maintaining financial security of the water supply system at the same time. Generally, conservation oriented rate structures comprise of an increasing block tariff, in which the price per unit increase step-wise with consumption, a seasonal tariff rate, in which the price per unit is higher during periods of peak summer demand or water shortages and/or water conservation surcharges, which impose additional tax on consumers to shape their long term water use behaviour. The most efficient conservation oriented tariff structure is one that provides financial incentives to consumers to reduce their inefficient or discretionary water usage, does not have negative impact on the revenue generation of the utility, avoid financial hardships on low-income households and delay investments on costly water supply expansion projects.

The design of a conservation oriented rate structures should take into account the local consumption patterns, the availability of water resources, the revenue requirements of the utility, and the conservation targets in the short and long term. The same utility or regulator can adopt different approaches to pricing under different circumstances. For instance, in its 2005 price review of Sydney Water, the Independent Pricing and Regulatory Tribunal (IPART) of New South Wales introduced a two tiered IBT structure (along with fixed costs) in view of its considerable potential in controlling demand in the face of drought conditions. However, in 2008, as water availability increased due to rising dam levels, increased recycling and construction of the desalination plant, the IBT structure was replaced by a single usage charge based on the LRMC of Sydney Water (Cox, 2010).

The price elasticity of demand for water can be a useful indicator of responsiveness of demand to changes in price and can offer insights into the variables that influence this responsiveness. In general, price elasticity is higher when outdoor discretionary uses comprise a significant percentage of the household's water use, when the initial rates are high and when consumers have greater potential to shift to water saving devices and adjust their behaviours. For the pricing structure to be effective in promoting change in water use behaviours, the water utility should educate customers about the rate structure and allow the customer to compare water use on his/her bill with average water use for his or her customer class, as well as his/her household water use for the last 12 months. The rate structure should be clearly explained on the water bill.

In the end, pricing is just one of the tools in the toolbox that a utility can use to shape the water demand of its consumers. Given the low price elasticity of demand for water in urban households, pricing can be effective when coupled with other demand management tools such as water restrictions, water saving devices and public awareness campaigns. Based on economic theories of cost-effectiveness and social considerations of equity and affordability, very complex tariff structures can be designed where separate rates exist for different categories and sub-categories of users. However, in practice, a complex tariff structure may achieve the same overall results at the cost of increased administrative complexity and consumer confusion. In the end, the best tariff structure is one that is simple and best suited to the local social, economic and environmental context.

ENDNOTES

[1]Section 3.1 and 3.2 of this chapter is based on my own research, published previously as: Hoque, Sonia F. and Dennis Wichelns (2013). State-of-the-art review: Designing urban water tariffs to recover costs and promote wise use. *International Journal of Water Resources Development*, **29**(3): 472–491.

[2]The marginal cost of supplying water is largely dependent on the capacity of large capital investments such as dams, and more recently desalination plants, and pipelines. Once this cost has been incurred, the marginal cost of supplying water is much lower than the average cost of supply. If prices are set to marginal cost the water utility may not fully recover its costs (Cox, 2010).

Chapter 4

Role of policy: Regulations and conservation programs

Many utilities across the world, especially in developed cities, have formulated policies to regulate the use of water within the household. These regulations can be in the form of mandatory restrictions involving certain water use activities during specific times of the day or week or in the form of voluntary conservation by providing consumers with informed choices. Regulations also include mandatory or voluntary labelling of water efficiency specifications on water fixtures and selling of fixtures below stated flow rates. In order to promote the adoption of water efficient fixtures by households, many governments and water agencies have also implemented public awareness campaigns and provided economic incentives. Many of these regulations or conservation programs have been largely successful in reducing water demand during periods of drought or in achieving long term decline in overall domestic water use. This chapter discusses the roles of these regulatory policies and programs in promoting water conservation in urban households. Section 4.1 discusses the costs and benefits of mandatory water use restrictions, as well as consumers' willingness to pay to avoid such command and control tools. Section 4.2 gives a brief overview of the policies used to promote water efficient devices, which have been dealt in greater detail in Chapter 5. Finally, Section 4.3 focuses on public awareness campaigns designed and implemented to raise people's knowledge about the importance of water as a scarce and precious natural resource. In this chapter, examples from Australia, the United States, Singapore and the Middle East have been provided to illustrate the effectiveness of policies in different settings.

4.1 RESTRICTION ON SPECIFIC WATER USE ACTIVITIES

Quantity rationing policies, such as mandatory restrictions on water consumption, are often used to reduce water demand in periods of short-run water shortages. The severity of the restrictions usually depends on the extent of shortage. Water use restrictions do not usually limit the absolute amount of water that can be consumed. Instead, they limit the types of water use activities, the methods in which water may be used as well as the timing of these activities. If properly enforced, these restrictions will eventually decrease the total amount of water used. Normally, restrictions are targeted towards non-discretionary water uses, such as watering gardens, filling swimming pools and washing hard surfaces. Box 4.1 discusses the water conservation measures implemented in Los Angeles to deal with water shortage.

BOX 4.1 WATER CONSERVATION IN LOS ANGELES – RESTRICTIONS, REBATES AND AWARENESS PROGRAMS

The Los Angeles Department of Water and Power (LADWP), the largest municipal utility in the United States, is responsible for supplying water to residents and businesses in Los Angeles city and surrounding areas. From 1980 to 2009, the population within LADWP's service area increased from 2.97 million to 4.1 million, representing an average annual growth rate of 1.3%. Within LADWP's service area, the largest category of demand is the residential sector, with single family and multi-family residents comprising 36% and 29% of the total demand respectively.

Los Angeles city historically receives water from five major sources: the Eastern Sierra Nevada watershed (via the Los Angeles Aqueduct, LAA); the Colorado River (via the Colorado River Aqueduct) and the Sacramento-San Joaquin Delta (via the State Water Project/California Aqueduct), which are purchased from the Metropolitan Water District of Southern California (MWD); local groundwater; and recycled water for industrial and irrigation purposes (Figure 4.1). Water from two of the supply sources, the LAA and MWD, is classified as imported because it is obtained from outside LADWP's service area.

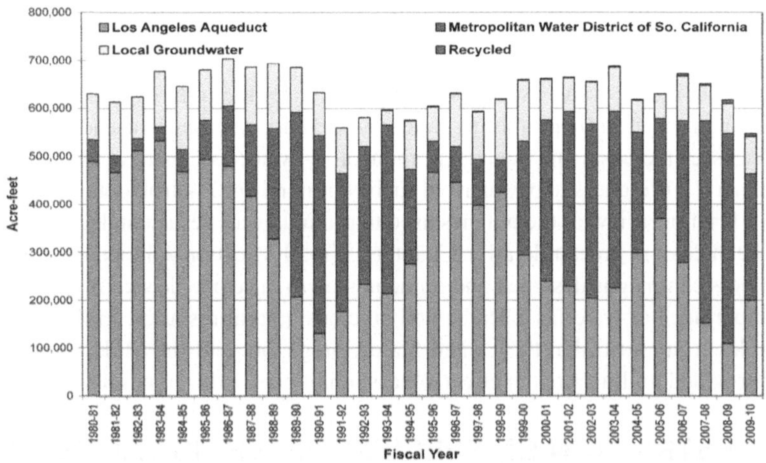

Figure 4.1 Water demand and sources of water supply in Los Angeles city (LADWP, 2010).

(Continued)

BOX 4.1 ***(Continued)***

During the 1980s, water demand in Los Angeles increased proportionately with population growth, reaching a peak of 863 million m^3 in 1986. This was followed by a five-year drought (1987–1992), during which a number of water conservation programs were launched by LADWP. These conservation initiatives, coupled with a boost in rainfall and economic recession, during 1992 and 1993, resulted in a 17% decline in total water demand in the following years. The average per capita water use decreased from 654 lpcd in 1989–90 to 601 lpcd in 1999–00 and finally to 442 lpcd in 2009–10. The per capita water consumption is still quite high compared to other cities because a significant proportion of the water supplied is used for outdoor purposes. Based on data collected from 2004 through 2007, for single-family and multi-family dwellings, 54% and 32% of domestic demand comprises outdoor water use.

While these sources have served the city for decades, the long-term sustainability of water supply is under threat due to several factors. Due to reallocation of water for environmental improvement in the Eastern Sierra, the city's water deliveries from LAA have dropped significantly over the last two decades. These environmental requirements include: the State Water Resources Control Board Mono Lake decision to permanently limit LADWP's ability to export water from the Mono Basin; implementation of the Owens Lake Dust Mitigation Project; re-watering of the Lower Owens River, and many other environmental restoration projects in the Owens Valley. Between 1995 and 2000, the city received 63% of its water from the Eastern Sierra through the LAA, which dropped to 34% during 2001 and 2004.

Moreover, in order to protect the delta smelt, there has been a Federal Court ruling that limits exports from the Sacramento-San Joaquin Delta by as much as one-third, which creates the possibility of future water allocations by MWD. Contamination of groundwater in the San Fernando Valley, the city's main source of groundwater, has also severely limited water available for pumping.

Given the increasing demand for water in urban areas and the constraints on historical water supply sources, Los Angeles has been implementing a multi-faceted demand management approach since the 1990s. This includes seasonal tiered water pricing (see Box 3.5), mandatory water use restrictions, rebates for installation of water saving devices and a number of awareness programs.

Installation of water efficient plumbing fixtures

Los Angeles has utilised ordinances as a tool to reduce water waste since 1988, beginning with the adoption of its first version of a plumbing retrofit ordinance. The ordinance mandated installation of conservation devices in all existing residential and commercial properties and installation of water-efficient landscaping in all new construction. Toilets were required to use less than 3.5 gallons per flush (gpf) (13 litres), urinals less than 1.5 gpf (6 litres), and showerheads less than 2.5 gallons per minute (gpm) (10 litres). Customers with three acres or more of turf were required to reduce water consumption by 10% from 1986 levels or face a 100% surcharge on their water bills. In 1998, the ordinance was amended, requiring the installation of ultra-low flush toilets and water saving showerheads in single family and multi-family residences. Los Angeles further increased its water efficiency mandates in 2009 with adoption of the Water Efficiency Requirements Ordinance (Table 4.1).

(Continued)

BOX 4.1 (*Continued*)

This ordinance establishes water efficiency requirements for new developments and renovations of existing buildings by requiring installation of high efficiency plumbing fixtures in all residential and commercial buildings.

Table 4.1 Specification under the Water Efficiency Requirements Ordinance (2009).

Devices	Requirement
High Efficiency Toilets	1.28 gallons per flush
Urinals	0.125 gallons per flush
Indoor Faucets	2.2 gallons per minute
Showerheads	2.0 gallons per minute
Domestic Dishwashers	5.8 gallons per cycle

Implementation of mandatory water restrictions

LADWP first adopted an Emergency Water Conservation Plan Ordinance in the early 1990s in response to drought conditions. In 2009, in response to a 3-year drought coinciding with an economic recession, LADWP has adopted two amendments expanding prohibited uses, increasing penalties for violating the ordinance, and modifying water conservation requirements. Five phases of water conservation are incorporated into the plan with prohibitions and water conservation measures steadily increasing by phase (Table 4.2). Regardless of water supply availability Phase I conservation requirements are in effect permanently unless a more stringent phase is in effect.

Customers of LADWP are liable to pay penalties if they do not abide by the water restrictions being enforced. Residential consumers with meters less than two inches, receive a written warning for their first violation. On second, third and fourth violations within the following 12 month period, an additional surcharge of USD 100, USD 200 and USD 300 will be added to the customer's water bill.

Rebates for water efficient plumbing fixtures

LADWP's past water conservation programs have assisted customers affected by the ordinances by offering free ULF toilets and showerheads, free installation of ULF toilets, showerheads and faucet aerators, as well as rebates for ULF toilets purchased and installed. Current water conservation programs, co-sponsored by MWD through the SoCal Water$mart Program for residential customers and the Save Water Save a Buck Program for Commercial, Industrial and Institutional (CII) customers, continue to assist customers in complying with ordinances and reducing overall water demands.

(*Continued*)

BOX 4.1 (*Continued*)

Table 4.2 Mandatory water restrictions in Los Angeles.

Phase I

- No use of a water hose to wash paved surfaces.
- No use of water to clean, fill, or maintain levels in decorative fountains, ponds, lakes or similar structures used for aesthetic purposes unless a recirculating system is used.
- No drinking water shall be served unless expressly requested in restaurants, hotels, cafes, cafeterias, or other public places where food is sold, served, or offered for sale.
- No leaks from any pipes or fixtures on a customer's premises; failure or refusal to fix leak in a timely manner shall subject the customer penalties for a prohibited use of water.
- No washing vehicles with a hose if the hose does not have a self-closing water shut-off device attached or the hose is allowed to run continuously while washing a vehicle.
- No irrigation during rain.
- No irrigation between 9 am and 4 pm, except for public and private golf courses and professional sports fields to maintain play areas and event schedules. System testing and repair is allowed if signage is displayed.
- All irrigation of landscape with potable water using spray head and bubblers shall be limited to no more than ten minutes per water day per station. All irrigation of landscape with potable water using standard rotors and multi-stream rotary heads shall be limited to no more than 15 minutes per cycle and up to 2 cycles per water day per station. Exempt from these restrictions are irrigation systems using very low-flow drip-type irrigation when no emitter produces more than 4 gallons of water per hour and micro-sprinklers using less than 14 gallons per hour.
- No watering or irrigation of any lawn, landscape, or other vegetated area shall occur in a manner that causes or allows excess or continuous water flow or runoff onto an adjoining sidewalk, driveway, street, gutter, or ditch.
- No installation of single-pass cooling systems shall be permitted in buildings requesting new water service.
- No installation of non-recirculating systems shall be permitted in new conveyor car wash and new commercial laundry systems.
- Operators of hotels and motels shall provide guests with the option of choosing not to have towels and linens laundered daily.
- No large landscape areas shall have irrigation systems without rain sensors that shut off the irrigation systems.

(*Continued*)

BOX 4.1 (*Continued*)

Table 4.2 Mandatory water restrictions in Los Angeles (*Continued*).

Phase II

- All prohibited uses in Phase 1 shall apply, except as provided.
- No landscape irrigation shall be permitted on any day other than Monday, Wednesday, or Friday for odd-numbered street address and Tuesday, Thursday, or Sunday for even-numbered street addresses. If a street address ends in ½ or any fraction it shall conform to the permitted uses for the last whole number in the address. For non-conserving nozzles (spray head sprinklers and bubblers) watering times shall be limited to no more than 8 minutes per watering day per station for a total of 24 minutes per week. For conserving nozzles (standard rotors and multi-stream rotary heads watering times shall be limited to no more than 15 minutes per cycle and up to two cycles per watering day per station for a total of 90 minutes per week.
- Irrigation of sports fields may deviate from non-watering days to maintain play areas and accommodate event schedules with written notice from LADWP. However, a customer must reduce overall monthly water use by LADWP's Board of Water and Power Commissioners adopted degree of shortage plus an additional 5% from the customer baseline water usage within 30 days.
- If written notice is received from LADWP, large landscape areas may deviate from the non-watering days if the following requirements are met: (1) approved weather-based irrigation controllers registered with LADWP; (2) Must reduce overall monthly water use by LADWP's Board adopted degree of shortage plus and additional 5% from the customer baseline within 30 days; (3) Must use recycled water if available.
- These restrictions do not apply to drip irrigation supplying water to a food source or to hand-held hose watering of vegetation, if the hose is equipped with a self-closing water shut-off device, which is allowed everyday during Phase II, except between the hours of 9am and 4pm.

Phase III

- All prohibited uses in Phases I and II shall apply, except as provided.
- No landscape irrigation shall be permitted on any day other than Monday for odd-numbered street address and Tuesday for even-numbered street addresses. If a street address ends in ½ or any fraction it shall conform to the permitted use for the last whole number in the address.
- No washing of vehicles allowed except at commercial car washes.
- No filling of residential swimming pools and spas with potable water.
- Irrigation of sports fields may deviate from non-watering days and be granted one additional watering days for a total of two watering days with written notice from LADWP. However, a customer reduce overall monthly water use by LADWP's Board of Water and Power Commissioners adopted degree of shortage plus an additional 10% from the customer baseline water usage within 30 days.

(*Continued*)

BOX 4.1 (*Continued*)

Table 4.2 Mandatory water restrictions in Los Angeles (*Continued*).

- If written notice is received from LADWP, large landscape areas may deviate from the non-watering days and be granted one extra day of watering for a total of 2 watering days if the following requirements are met: (1) approved weather-based irrigation controllers registered with LADWP; (2) Must reduce overall monthly water use by LADWP's Board adopted degree of shortage plus and additional 10% from the customer baseline within 30 days; (3) Must use recycled water if available.
- These restrictions do not apply to drip irrigation supplying water to a food source or to hand-held hose watering of vegetation, if the hose is equipped with a self-closing water shut-off device, which is allowed everyday during Phase III, except between the hours of 9 am and 4 pm.

Phase IV

- All prohibited uses in Phases I, II, and III shall apply, except as provided.
- No landscape irrigation is allowed.

Phase V

- All prohibited uses in Phases I, II, III, and IV shall apply, except as provided.
- The LADWP Board of Water and Power Commissioners is authorised to implement additional water prohibitions based on the water supply situation.

Source: LADWP (2010).

Awareness programs

Besides active awareness and support programs that included universal metering, assessment of volumetric sewer charges, and a conservation rate structure, LADWP also implemented passive components such as providing educational materials for schools, community and customer presentations, maintaining a conservation hotline, and a wide range of information distributed through customer bills, advertising in public venues, LADWP's website, and direct mail. By raising awareness on water consumption, providing visibility to water conservation program, and encouraging community involvement, these measures form the foundation for the conservation movement.

Funding and cost effectiveness

The cost range of conservation rebates, incentives, and hardware installation programs ranges from approximately USD 75 to USD 900 per acre-feet based on current LADWP conservation programs. Compared to other alternatives, investment in water conservation seems to be most cost-effectiveness for LADWP. The average unit cost is USD 563 per acre-feet for LAA water, USD 527–869 per acre-feet for MWD water, USD 600–1500 per acre-feet for recycled water and USD 1300–2000 per acre-feet for desalination. Water conservation programs are primarily funded through water rates. MWD offers both commercial and residential rebates to member agency customers that install specified conservation devices. Other sources of fund include co-funding from outside agencies and grants from the state. More than USD 200 million has been invested in water conservation since 1991.

Source: LADWP (2010) and Villaraigosa (2008).

4.1.1 Costs of restrictions

Although restrictions do not usually impart direct financial costs on households and communities, there are several hidden costs associated with this type of demand management approach.

4.1.1.1 Costs to households

By forcing all consumers to reduce water use in the same way, mandatory restrictions deny households the opportunity to use water in ways that are most valuable to them. For instance, avid gardeners or families with children may place higher value in maintaining a green landscape or indulging in water related leisure activities, than those living in high rise apartment buildings. This leads to an inefficient allocation of water by not taking into account individuals' willingness to pay for certain water use activities. Moreover, specifying the times and means by which gardens can be watered increases the time and effort needed to do these activities. Some people may also find taking longer showers or swimming in a pool as de-stressing. Lack of regular maintenance, due to restricted water use, can degrade the amenity values of pools and gardens and degrade these assets (Colmar Brunton Social Research, 2008).

To compensate for restricted times of use, there can be 'over watering' of gardens during the allowable watering times (Brennan *et al.*, 2007). The complexity of the arrangements, such as which days of the week and times of the day watering is allowed, and when odd/even house numbers are permitted to use water, can create confusion among consumers. Households may also end up spending more on purchasing and installing private water storage equipment, such as rainwater tanks, and watering systems, such as hand-held hoses, to comply with changes.

4.1.1.2 Costs to the community

Mandatory restrictions can reduce community welfare through a loss of amenity associated with less green open space, including unwatered council parks and reduced access to community sport and recreational facilities. This loss of amenity can lead to other social problems including increased health issues, such as depression and obesity. Poorer quality sporting fields can also lead to risk of injury from sporting activities (Colmar Brunton Social Research, 2008).

Mandatory restrictions can also have environmental impacts through a loss of green open space, including reduced cooling effects on buildings, requiring greater energy consumption, diminished urban stormwater management, as green open spaces slow runoff after rainfall and filter pollutants, and distorted soil structure and soil erosion (CRCIF, 2008). A lack of green open space can also reduce property values, and cause damage to buildings, other structures and pipes through cracking. It has been argued that mandatory restrictions can also lead to a reduction in social cohesion arising from households being encouraged to report neighbours that do not comply (Institute for Sustainable Futures and ACIL Tasman, 2009).

4.1.1.3 Costs to utilities and local governments

Administration of restrictions imparts financial costs on governments and utilities. Restrictions require advertising campaigns to communicate the rules under different levels of restrictions. Where mandatory, there are also monitoring and enforcement costs involved (CIE, 2008). While using prices to curb water demand can achieve the dual objective of reducing water use while maintaining revenues, restricting water use significantly lowers revenues for water utilities by decreasing the volume of water sold.

4.1.2 Benefits of restrictions

The effectiveness of using prices to control water demand is largely determined by the price elasticity of demand (see Chapter 3). Moreover, changes in prices do not have immediate effects as consumers need time to react and change their water use behaviours. Comparatively, restrictions can be more effective, especially in the short run, if they are enforced properly.

4.1.2.1 Reduction in water consumption

Water restrictions have been effective in reducing the short-term and long-term demand for water in many cities. Restricting outdoor water use was a key element in securing greater Sydney's water supply during the millennium drought in 2000s. Between the introduction of Level 1 restrictions in October 2003 and the introduction of permanent Water Wise Rules to replace Level 3 restrictions in June 2009, restrictions saved an estimated 575 billion litres, more than the amount of water used in greater Sydney in 2009–10 (Productivity Commission, 2011). According to the Australian Productivity Commission, mandatory water restrictions are of most benefit when a quick response is needed. During emergencies, the price mechanism can be too slow to yield the change in demand required.

4.1.2.2 Community support

In certain cases, mandatory water restrictions are highly valued by the community. Water saving activities give individuals a sense of community spirit and solidarity by working together to achieve a common purpose. They are seen by many as an equitable tool for dealing with water shortages because the losses from rationing a shortage of water are shared equally by all households (Productivity Commission, 2011). It is sometimes argued that such restrictions are more equitable than using pricing tools. When prices are used to manage demand, those that can afford to buy more have access to larger amounts of water than poorer households (Colmar Brunton Social Research, 2008).

4.1.2.3 Environmental benefits

A decrease in overall water consumption is beneficial to the environment as less water is extracted from surface and groundwater sources, which in turn, leads to less wastewater being discharged. In addition, there are also indirect benefits, resulting from reduction in energy used in treating, pumping and heating water and the corresponding decline in carbon emissions.

4.1.3 Consumers' willingness to pay to avoid restrictions

Although outdoor water restrictions have been effective in reducing urban water demand during periods of shortfall in supply, consumers may not be willing to comply with such mandatory restrictions that affect their way of life. A number of studies have used econometric analysis of panel data obtained by combining household water consumption data from water utilities, with household characteristics data derived from surveys or secondary data, in order to assess the efficacy of different types of drought management policies and their distributional implications (e.g., Renwick & Archibald, 1998; Renwick & Green, 2000; Schuck & Profit, 2004; Garcia-Valinas, 2006). Other studies have examined whether there is any evidence of a demand response to general conservation programs which may include policies such as advertising, retrofitting indoor appliances, as well as outdoor restrictions (e.g., Nieswiadomy, 1992). A number of studies have also investigated individuals' preferences in relation to drought water restrictions using the stated preference techniques of choice experiments and contingent valuation (Howe & Smith, 1994; Griffin & Mjelde, 2000; Gordon *et al.*, 2001; Koss & Khawaja, 2001; Hensher *et al.*, 2006). Griffin and Mjelde (2000) and Koss and Khawaja (2001) employed contingent valuation to evaluate customers' willingness to pay (WTP) to avoid water restrictions in the USA. Griffin and Mjelde (2000) found that respondents would be willing to pay an average of USD 25.3–34.4 per month to avoid an occurrence of water restrictions (in 1997 USD), while Koss and Khawaja (2001) found WTP values of USD 11.7–16.9 per month (in 1993 USD).

Gordon *et al.* (2001) used a choice experiment to investigate the preferences of residents' in Canberra toward water restrictions. Their study indicated that consumers were willing to pay an extra AUD 150 per year (in 1997 AUD) on their water bill for a more voluntary based demand management approach rather than mandatory restrictions. In another choice experiment study, Hensher *et al.* (2006) studied households' WTP to avoid drought water restrictions in Canberra. The study suggests that, on average, residential customers were willing to pay AUD 239 per year on top of their water bill to move from a situation with continuous restrictions to a situation with virtually no chance of restrictions.

Brennan *et al.* (2007) showed that the preferences towards lawn greenness, and the time cost, or disutility associated with hand-held watering, influence the extent to which a sprinkler restrictions policy is effective. The authors estimated that the household welfare costs of a sprinkler restriction appear to be less than AUD 100 per season when mild (two days per week) sprinkler restrictions are in place, and may range between AUD 347 and AUD 870 per season when a complete sprinkler ban is in place.

To understand the welfare costs associated with urban water restrictions in six Australian cities, Cooper *et al.* (2011) conducted a multiple-bounded discrete choice contingent valuation experiment to identify households' willingness to pay to avoid urban water restrictions. According to their results, respondents with a lawn were willing to pay AUD 157–267 per year to avoid restrictions, while those without a lawn were willing to pay AUD 113–223 per year. Similarly, the willingness to pay to avoid restrictions ranged from AUD 119–229 for low income households and AUD 182–292 for high income ones. Notwithstanding the variance in the contexts of these studies, the results suggest that water restrictions impose a utility loss on households and they are prepared to pay to avoid that loss.

4.2 PROMOTION OF WATER-SAVING DEVICES

In an urban household, water is used via bathroom and kitchen appliances, which have fixed flow rates or water consumption under different settings. The terms 'water saving devices' or 'water efficient fixtures' are used to refer to appliances which require less water for achieving the desired task, compared to their older counterparts. In order to promote the use of water saving devices such as faucets, showerheads, toilets and washing machines, some countries have passed legislations to mandate the installation of water-saving appliances in houses built after a certain year, while others have made it mandatory for water appliance manufacturers to label their products with water efficiency specifications so that consumers can make informed choices (see Section 5.3). Some governments and utilities also give rebates to consumers to retrofit their old appliances with water-saving devices (see Section 5.4). The role of water saving devices in reducing water consumption has been discussed elaborately in Chapter 5. In this chapter, the plumbing codes implemented in the United States are described in Box 4.2.

BOX 4.2 PLUMBING CODES IN THE UNITED STATES

In the 1970s, due to aging water and wastewater utility infrastructure and shrinking water supply, utilities encouraged manufacturers to design and sell toilets that could flush less water than the five to seven gallons per flush (gpf) common in that time period (Osann & Young, 1998). By the late 1980s, all major toilet manufacturers produced toilet models that flushed 1.6 gpf, and, by the early 1990s, a coalition of plumbing product manufacturers, water and wastewater utilities, environmental organisations, and other interested parties petitioned Congress to enact national efficiency standards for plumbing products (Osann & Young, 1998). Congress did so in 1992 through the Energy Policy and Conservation Act.

The Energy Policy Act enacted in 1992 outlines mandatory standards for water appliances such as replacing the 3.5 gpf (gallons per flush) toilet with a new 1.6 gpf or 6 litres per flush maximum standard for all new toilets (Table 4.3). In 1994, the act was extended to showerheads and faucets sold in the USA, which should not release more than 2.5 and 2.2 gallons of water per minute (gpm) respectively. Moreover, in 1994 the American Water Works Association (AWWA) established a clearing house called WaterWiser to support water conservation professionals and the general public in using water more efficiently. In 2006, the U.S. Environmental Protection Agency (USEPA) launched the WaterSense program to encourage water efficiency beyond the standards set by the Energy Policy Act by labelling water appliances. WaterSense partners with manufacturers, retailers and distributors, and utilities to bring WaterSense labelled products to the marketplace and make it easy to purchase high-performing, water-efficient products (Figure 4.2).

Table 4.3 National efficiency standards and specifications for residential and commercial water-using fixtures and appliances.

Fixtures and appliances	EPAct 1992, EPAct 2005 standard	WaterSense standard
Residential toilets	1.6 gpf	1.28 gpf with at least 350 g waste removal
Residential bathroom faucets	2.2 gpm at 60 psi	1.5 gpm at 60 psi (no less than 0.8 gpm at 20 psi)
Residential showerheads	2.5 gpm at 80 psi	2.0 gpm at 80 psi (effective from February 2010)
Urinals	1.0 gpf	No specification
Residential cloth washers	No specification	Water factor $\leq$8.0 gal/cycle/ft^3
Residential dishwashers	No specification	No specification

Source: U.S. EPA (2008, 2010).

Some states had set minimum efficiency levels for plumbing products prior to the federal action, but the federal standard pre-empted state standards. A clause in the

(Continued)

BOX 4.2 (*Continued*)

law stated that, if the standards did not change in five years, the pre-emption would be waived. This waiver did not happen for years, causing uncertainty as to whether state or local jurisdictions could set stricter standards (North Carolina Department of Environment and Natural Resources, 2009). In December 2010, the federal government finally released a waiver of pre-emption that would allow states to set higher efficiency standards for these products (Department of Energy, 2010).

Figure 4.2 WaterSense product labelling in the United States.

Some states, including California (Assembly Bill 715-Chapter 499) and Texas (HB No. 2667), have implemented stricter plumbing codes. The California plumbing code requires 100% of toilets and urinals (other than blow-out urinals) sold or installed in California, on or after January 1, 2014, to be high-efficiency (maximum of 1.28 gpf for high-efficiency toilets – HETs – and 0.5 gallons per flush for high-efficiency urinals – HEUs). In addition, the law requires that non-water urinals be approved for sale and installation in California.

4.3 PUBLIC AWARENESS CAMPAIGNS

Water conservation awareness involves understanding the necessity to use water efficiently at all stages, that will in turn promote attitudes and behaviours in relation to water use. One of the most critical components of a program is a robust education and outreach program that reaches water provider employees, school children, and adults.

Water providers are responsible for planning and managing water supply. It is imperative that they understand that freshwater is a finite natural resource that plays a significant role in a country's overall economic development. Water providers can be educated through seminars and workshops, and selected staff can be sent to technical and professional development courses in countries with a strong conservation culture (UNESCAP, 2002). For example, the Temasek Foundation Water Leadership Programme, developed in collaboration between

Temasek Foundation, the Lee Kuan Yew School of Public Policy at the National University of Singapore and PUB, Singapore's national water agency, offers a two-week executive programme for water utilities managers in Asian countries, where there is constant pressure of increasing water supply to meet the growing demand. The program enables participants to learn about the best practices in water governance, get practical guidance on effectively managing water utilities, and develop strategies to deal with current and future challenges (NUS, 2013).

Water users refer to the general population who are the customers of water utilities. Domestic customers can be given practical water-saving tips at home, details of water-saving devices and information on the real cost of water services, through their water bills or other communication tools. Large customers, for example water intensive industries, can be offered water audits for free (UNESCAP, 2002).

Educating children and students serves two purposes: firstly, it inculcates an attitude for water conservation among the future society; and secondly, it raises awareness in the present society when children share their learnings with their families. Education materials can be integrated with the curricula in schools, colleges and universities, and young children can be taught through hands-on interactive sessions. Effective measures include jigsaw puzzles, board games, quizzes, audio or video tapes and stories with popular characters. Teachers should undergo preparatory training and be provided with a curriculum guide, background information, student worksheets and other material for class work. Water utilities can establish student visitor centres and mobile exhibitions, organise educational visits to water facilities and provide guest speakers to schools (UNESCAP, 2002).

Education of the public at large, municipal officials and water suppliers is crucial to generating an understanding of the issues, and creating acceptance to the implementation of water conservation efforts. It is important to provide to the public the basic understanding of the water supply situation, the cost of delivering water, the needs for water conservation and the roles that people can play in saving water resources. Box 4.3 discusses the various types of awareness raising tools used in Singapore over the last three decades to educate the general public, both adults and children, about the importance of water conservation. This is followed by Box 4.4 which highlights how religious norms and beliefs have formed the basis of public awareness programs in some parts of the world.

Traditionally, water conservation information campaigns have been implemented during times of severe drought, either independently or in association with other demand management tools such as restrictions and price increases. In recent years, these campaigns are conducted on a more regular basis to inculcate a water conservation culture among citizens. There has been polarisation of opinions with respect to the effectiveness of these campaigns in reducing water use and changing attitudes and behaviours. On one hand, while awareness campaigns are politically acceptable and promote socially responsible behaviours, on the other hand, such campaigns may not be cost effective and have only temporary results.

BOX 4.3 RAISING PUBLIC AWARENESS AND PROMOTING PUBLIC ENGAGEMENT IN WATER CONSERVATION IN SINGAPORE

In Singapore, lack of adequate local freshwater resources, dependence on imported water from Malaysia and increasing water consumption due to population and economic growth have compelled the government to look for alternative sources of water supply and promote water conservation (see Box 1.2). While pricing structures have been revised periodically to address cost recovery and promote wise use (see Box 3.1) and regulatory policies have been implemented to prohibit water pollution and using water efficient devices, the government also launched many public awareness campaigns, since the 1960s, to communicate the importance of water as a scarce and precious resource as well as to actively engage the community in water management and instil a sense of ownership.

Following Singapore's independence in 1965, increase in overall water consumption was seen as a positive indicator of economic growth and rising living standards. However, severe droughts in 1971 and 1976, coupled with continued increase in water consumption in the domestic and non-domestic sectors, led to concerns about conserving water resources. While initial efforts to promote water conservation involved general information dissemination and awareness raising through education, in later years the Public Utilities Board (PUB) became more actively involved by conducting site visits and liasioning with industrial users and households to provide overall support on reducing water consumption.

Public awareness campaigns on environmental matters, including water resources, have been initiated in Singapore since 1968. In 1972, PUB launched its first large-scale consumer-oriented campaign, named 'Water is Precious', to make the public aware of the growing importance of water and to instil water saving habits. Numerous community activities were organised highlighting the practical dos and don'ts for conserving water and press, radio and television publicised extensively the campaign among the public.

In 1976, after another serious drought, PUB published appeals to large water users, such as hotels, coffee shops and laundries, to avoid wasting water (PUB, 1976). Slogans such as 'Don't wait till the last DROP – Save water now' were displayed in public places, disseminating the need to save water. The 'Water is Precious' exhibition was revived and ministries, agencies and associations including the Prime Minister's Office, the Ministry of Culture and the PA collaborated, worked and toured 12 community centres together to reinforce the importance of water conservation.

With the new slogan 'Adapt, Innovate and Prosper', in 1978, PUB started to change its approach from passively educating the public to actively involving them in understanding their roles and responsibilities. This included the implementation of multiple initiatives, such as the 'Meter Reading Contest' and tours to water treatment works and power stations for new employees, students, organised groups and overseas visitors to learn about the work of the utilities. A year later, to bring people closer to water, two of the reservoirs (Kranji and Upper Peirce) were opened up for recreational uses (|PUB, 1978, 1979).

In 1981, a one-month long 'Let's Not Waste Precious Water' campaign was launched and PUB officials participated in talk shows in schools, colleges and vocational institutions to encourage water conservation attitudes (PUB, 1981). In the same year,

(Continued)

BOX 4.3 (*Continued*)

the 'Water Conservation Unit' was established to work on water conservation in the domestic and non-domestic sectors and advice on the most suitable policies to be adopted. Officers from the unit inspected large industrial and commercial consumers and persuaded them to adopt water conservation measures, such as the installation of self-closing delayed action taps and constant flow regulators. In order to highlight the water saving achievements of large consumers, the government announced a 50% investment tax allowance for water conservation equipment for those industries that had substantially reduced their water consumption (PUB, 1983).

In 1986, the Singapore Civil Defence Force, PUB, several grassroots organisations and approximately 3500 households participated in an 'emergency' water exercise (PUB, 1986). This activity was meant to target the population born after independence and that had never experienced a water crisis. In order to have a greater impact on the younger generation, a water conservation course was introduced at secondary school level (PUB, 1987).

The launch of the 'Singapore Green Plan', in 1992, marked the beginning of 'public consultations' as an engagement tool in environmental issues in Singapore and a paradigm shift in Singapore's water education from a more top-down mandatory approach to a greater inclusion of participatory bottom up and voluntary measures. In 1995, to remind Singaporeans about the importance of water, an island-wide water rationing exercise was conducted involving 30,000 households in 20 constituencies. During this period, water supply was interrupted for 14 hours each day (PUB, 1995). Other campaigns, such as 'Use water wisely' and 'Turn it off, Don't use water like there's no tomorrow', were conducted to emphasise the value of water and complement the rapid changes in water and wastewater tariffs in the late 1990s.

During the mid-2000s, PUB planned to disseminate messages in more subtle and emotional ways but with more permanent impacts. By bringing people closer to water bodies and allowing them to enjoy water-related activities, PUB aimed to build people's affinity to water so that they would gradually take ownership of the water bodies and stewardship in their conservation. In 2004, activities encouraging the public to enjoy water and develop a relationship with it were introduced in what became the '3P' ('People, Public and Private') approach. In 2006, PUB launched the Active, Beautiful, Clean Waters (ABC Waters) Programme as a strategic initiative to improve the quality of water and life by harnessing the full potential of our water bodies. By integrating the drains, canals and reservoirs with the surrounding environment in a holistic way, the ABC Waters Programme aims to create beautiful and clean streams, rivers, and lakes with postcard-pretty community spaces for all to enjoy. Till June 2013, there have been 20 successful ABC Waters projects around the island and over 100 ABC Waters proposals have been identified for implementation over the next decade.

In 2006, PUB embarked on the '10-Litre Challenge' programme to challenge every Singaporean to save 10 litres of water a day by implementing seven water saving tips displayed on PUB's website (visit http://tenlitres.com.sg/main.htm). In 2011, with its tagline 'Every drop counts, Use only what you need', PUB focused on two of the most water consuming activities at home: shower and washing in kitchen sink, which takes up 29% and 22% of an average family's monthly water consumption respectively. By shortening

(*Continued*)

BOX 4.3 (*Continued*)

the duration of water usage, people can save 9 litres of water per minute in shower and 8 litres of water per minute in the kitchen sink (PUB, 2011a).

To put a 'face' to Singapore's water management issues, the 'Water Wally' – the official mascot of PUB, was launched in 2005 (Figure 4.3). The 'Water Wally' has been extensively featured in T-shirts, souvenirs, NEWater bottles, water saving kits, posters and other publicity materials. In 2009, a nine-episode short animated series featuring Water Wally entitled 'The Adventures of Water Wally' was telecast on TV channels (visit http://www.pub.gov.sg/waterwally/). In 2011 and 2012, a series of television advertisements focusing on taking shorter showers, not washing dishes under a running tap, installing thimbles in taps and showerheads and benefits of water efficient washing machine were telecasted in the four official languages of Singapore. In 2011, as part of the effort to get the students from primary schools to share their views on water conservation, PUB organised a video competition, the winners of which were selected by a panel of judges and online voting.

Figure 4.3 Water Wally balloons to welcome visitors at the opening of Marina Barrage (Singapore's 15th reservoir) in 2008 (AsiaisGreen, 2008).

In 2013, the mascot has been featured in a promotional video 'The Water Wally Shower Dance' to urge children and adults to limit their showers to five minutes. A timer and an activity booklet have been given to school students to track their shower timings for a week. These students have also taken on the roles of junior water advocates by encouraging their parents to take shorter showers and spread water conservation tips to their neighbours. Known as 'Time to Save', this programme is run in conjunction with the new Primary 3 Social Studies Syllabus where students are taught water conservation; in particular by taking shorter showers. This campaign also involved the Water Wally displaying conservation messages to commuters in MRTs (mass rapid transit). Like video competition in 2011, PUB's 2013 initiative also involved a five month competition titled, 'My Take on Water'. The goal of this competition was to invite the public to share their take on water and how it has transformed or touched their lives (PUB, 2013a).

Source: Tortajada and Joshi (2013); PUB Annual Reports and website.

BOX 4.4 WATER CONSERVATION AND PUBLIC AWARENESS CAMPAIGNS IN ISLAM

In Islam, the relationship between humans and water is part of the daily social existence. This is well documented in the Quran and the *Sunnah* (Statements or practices undertaken or approved by the Prophet considered as legally binding precedents). Water is described as a gift by God so that humanity can benefit from it. This gift is the proof of the existence and uniqueness of Allah (Sura of the Ant, 27, 60). Water is also a symbol of resurrection since paradise is always described as a place with rivers flowing and florid vegetation (Sura of the Bee 16, 30–31). Several verses of the Quran also focus on the importance of water for purification and personal cleanliness (Sura of Booty 8,11).

Since religion exerts a very big influence on society and because water plays an extremely pivotal role within Islam, awareness campaigns and water conservation programmes based on religious principles have proved very useful, beneficial and cost effective in different areas of the Muslim world. The main reason why Islam and Islamic principles have been used as a vehicle in several awareness campaigns in predominantly Muslim states since the 1990s is because they can easily reach all strata of society.

One of the first steps was to use mosques as a platform for the campaign because of their accessible location and conspicuous number in every urban centre and because of the aggregative role they play within Muslim society. In Afghanistan, Jordan, Palestine, Egypt and also in some Gulf countries, Imams have been properly trained on water issues so that during the *Khutbah*, the Friday sermon, they could instruct all the believers on new techniques for water conservation or just draw their attention on water scarcity and the means to alleviate the problem at least in their domestic environment. Imams would focus on Quranic and Sharaitic precepts thus providing a solid basis for their speech and would then shift to more up-to-date problems such as scarcity afflicting the surrounding environment or more modern subjects like wastewater reuse.

Posters, leaflets, booklets and stickers have also been used to promote awareness on water issues in the Jordan River area (Palestine, Jordan and Syria). This way people could be reached in the streets or in their households, thus not confining the campaigns to a limited environment. Most of these posters use a religious terminology and imagery stressing the fact that saving water is a religious duty. An interesting example that shows how Islam and water issues are greatly interlocked is the fact that the water awareness campaign by WEDO, a Palestinian NGO, has been named 'Zam Zam' like the water spring in Mecca, where Hagar and Ishmael quenched their thirst after wandering in the desert for days on end. Any Muslim, upon hearing the name Zam Zam would associate it with water thus making the main focus of this campaign easily understandable.

Booklets have often been released by official organisations and institutions giving instructions on water conservation. There is a series of pamphlets published by the World Health Organisation on Health Education through Religion in the Eastern Mediterranean Region, a prominently Muslim area. The pamphlet on 'Water and Sanitation in Islam' (Al-Husseini Al Sheikh, 1996) makes parallels between correct behaviours in water use and conservation and Hadiths by the Prophet on the same subjects. The Jordanian Ministry of Water has also released various pamphlets on water management, all of them ending with several Quranic verses pertaining to water.

Source: Atallah et al. (1999) and Al-Husseini Al Sheikh (1996).

It is often difficult to quantify the impacts of information campaigns on water consumption patterns and pro-environmental attitudes, especially when these programs are launched in conjunction to other demand management tools. Previously, water savings from non-price programs were estimated from engineering calculation of expected reductions or based on comparisons of a utility's water demand before and after the program was implemented. In most cases, these resulted in overestimations as other factors affecting demand were not controlled for (Michelson *et al.*, 1999). Syme *et al.* (2000) reviewed the literature on the effectiveness of public awareness campaigns on water use and attitudes. The results of the few studies that have been conducted in the past, show that findings are highly depended on the methodologies used. The authors identified two types of approaches used by researchers: (1) narrative approach, in which the researcher employs information from case studies to assess the outcomes of an information campaign; and (2) statistical approach, in which researchers use regression models or other statistical tools to estimate the contribution of a number of variables (e.g., price changes, campaigns, weather) to the overall water savings. Other relatively less used approaches are quasi-experimental and experimental methods.

Among the studies that used the narrative approach, the Century Research Corporation (1972) analysed the effectiveness of winter conservation campaigns during droughts in 17 towns and cities in the United States and concluded that if information campaigns are commenced at the early stages of the drought, they can result in an overall reduction of 15–30% in water use. Similarly, information campaigns led to about 20% decrease in consumption during the 1976 drought in Britain (Blackburn, 1978), 15–20% reduction during the mid- to late 1970s droughts in the United States (Gilbert, 1978), and upto 28% savings in residential water use during the 1986 drought California (Berk *et al.*, 1993).

Syme *et al.* (2000) noted that sophisticated regression based studies are associated with problems of multi-collinearity of dependent variables, that is, it is difficult to evaluate the effectiveness of individual mechanisms when multiple demand management tools are employed, especially during short-term droughts. Hence, the studies produce general information about which components were successful in relation to others, but do not provide precise results that can contribute to planning the best mix of programs. For example, Michelson *et al.* (1999) used cross-sectional monthly time series domestic water demand model to examine the effects of non-price programs (such as information campaigns, school education programs, retrofit programs and regulations) in seven cities in California, New Mexico, and Colorado from 1984 to 1995. While the study found that as a whole these programs result in significant reductions ranging between 1.1–4%, the authors were unable to distinguish the effects of individual programs due to lack of information. The authors also observed that non-price programs exhibited diminishing returns, indicating that the effectiveness of a single program declined as more and more programs were implemented in the same city. The papers by Nieswiadomy (1992), Wang *et al.* (1999), Maki *et al.* (1978), Mercer and Morgan (1980) and Bruvold

(1979) are examples of some noteworthy studies that used statistical methods to investigate the relative impacts of various demand management tools.

Experimental research designed to promote water conservation through information and feedback strategies have been very limited. Geller *et al.* (1983) studied the effects on educational, behavioural and engineering interventions on water conservation in 129 households in Virginia. The educational approach involved distributing handbooks containing information on the problems of wasting water, the relationship between water and energy consumption and the ways of saving water at home. The behavioural approach involved daily written information feedback (on previous day's water consumption) with social commendation (smiling or frowning faces to highlight increase or decrease in consumption), and weekly summary charts of water consumption during baseline and treatment phase. The engineering approach involved installing a set of water conservation devices, with information on how these functioned. The study found that only the installation of low flow devices significantly reduced water use (although lower than expected), while the educational and behavioural interventions were ineffective in reducing consumption. According to the authors, the ineffectiveness resulted from few factors: firstly, the low price and inefficient rate structure (decreasing block tariff in this case) provided no economic incentive for conservation; secondly, the information feedback was read by only one person within the household and thus had no effect on other members; and finally, people seemed to have increased their water consumption in response to presumed savings from the low flow devices.

Recently, Fielding *et al.* (2012) investigated the effectiveness of a range of possible intervention strategies to reduce household water use in South East Queensland. A sample of 221 households was divided into four groups – three different intervention types and one control group – and their ongoing daily water consumption was monitored before, during and after the intervention. In the information only condition, households received general advice about how they could save water; in the descriptive norm condition, households received the general information about water saving along with information about other 'low water use households' that used these same behaviours; in the water end-use feedback condition, households received the general water saving tips along with tailored information (derived from smart meters) specifying where water was being used in their own household; and the control group received no general water saving information. Using longitudinal modelling, the study found that the three interventions all resulted in significant reductions in household water consumption relative to the control group.

4.4 CONCLUSION

Regulatory policies and water conservation programmes, such as outdoor watering restrictions, rebates on and labelling of water saving devices, and public awareness campaigns, are the main tools for managing demand using non-price approaches. Outdoor water restrictions are usually implemented in phases with varying degrees of limits and the onset of the phases are linked to the severity of water shortfall in reservoirs. While there are debates that outdoor restrictions impart a disproportionate burden on households with higher priority for such activities (see Chapter 6) and exert indirect costs in terms of monitoring and enforcement, they have immediate short term effects and are more politically acceptable than price increases. In case of water saving devices, there can be legislation that mandates the sale and installation of product with certain specifications or legislation that requires manufacturers of water fixtures to display water use information on the products. Water saving devices have penetrated the market since 1990s and have opened a new avenue for competition among manufacturers. Finally, it is imperative that the water providers and consumers understand that freshwater is a limited resource and it should be used efficiently for the common good of the society.

Chapter 5

Role of techonology: Water saving devices

The level of water consumption in an urban household is determined by a number of socio-economic, demographic and psychological factors such as household size, household income, dwelling type and ownership, water tariffs, regulatory policies, attitudes and behaviours, climate and technology (as discussed in Chapter 2). This chapter particularly focuses on the technological aspect, that is, the specifications or flow rates of various water appliances installed in a house. The volume of water required per flush, the flow rate of a faucet or showerhead and the water required by a washing machine or dishwasher per cycle are, in fact, the primary determinants of water consumption in a household. As these devices are usually fitted during the construction phase of new houses, they have the greatest potential for water conservation compared to modification of consumer behaviour during the occupier phase (Figure 5.1). For older buildings, the challenge is greater, as retrofitting old appliances with newer water saving devices can be more expensive and the owner/occupier might be reluctant to undergo additional hassles.

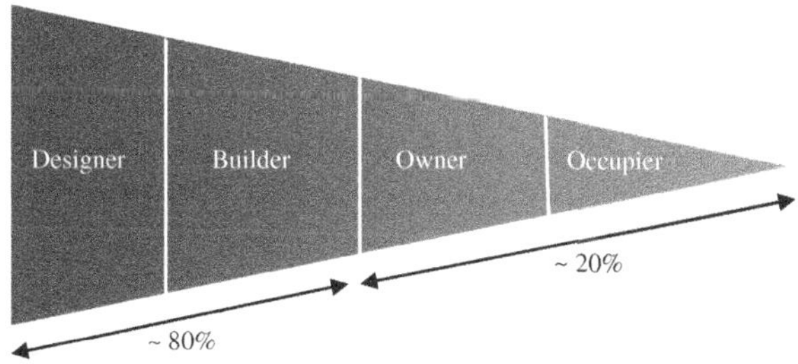

Figure 5.1 Potential for water conservation at various phases of the lifecycle of a house.

A wide range of regulatory policies and water conservation programs have been implemented by governments and water utilities to promote the installation of water saving appliances (or water efficient fixtures) within the household (as discussed in Chapter 3). These devices, which include faucets, toilets, showerheads, washing machines and dishwashers, use significantly less water than their older counterparts, without sacrificing performance or user satisfaction. Water efficient fixtures are widely available in the market both in developed and developing countries. Compared to other mechanisms of curbing water usage, such as through prices or mandatory restrictions, installation of water saving devices in new buildings and retrofitting conventional appliances in older buildings are much more effective and certain means of reducing water consumption. Once installed, these devices will continue saving water over their lifetime, provided there are no behavioural changes among the consumers. For instance, a showerhead with a lower flow rate will save significant amounts of water each day, given that the user does not start taking longer showers after its installation.

Section 5.1 provides an overview of the various types of water-saving devices available and their technical specifications. Section 5.2 discusses the major factors which promote the adoption or installation of water saving devices in a household, with reference to research studies conducted by various authors. As case studies, this chapter contains boxes on water conservation using water efficient devices in United States (Box 5.1), water efficiency labelling scheme (WELS) in Singapore (Box 5.2), costs and benefits of WELS in Australia (Box 5.3) and the effect of water efficient devices on diurnal demand patterns in Queensland, Australia (Box 5.4).

5.1 OVERVIEW OF WATER SAVING DEVICES

5.1.1 Toilets

Toilets are by far the main source of water use in the home, accounting for nearly 30% of an average home's indoor water consumption. Although the number of flushes per day depends on the extent of use, the volume of water used per flush exclusively depends on the technology of flushing. After their initial launch, many people disliked the low-flow toilets because they had to flush the toilets twice to achieve their desired task. Over the past two decades, newer types of water efficient toilets have been designed, which deliver water savings as well as high performance.

Single flush toilets are the most common and cheapest forms of toilets which use the force of gravity to transfer water from the tank to the bowl. Prior to 1980, these toilets used 5–7 gallons per flush (gpf) and after 1980, 3.5 gpf (13.2 litres) toilets became the standard. The US federal Energy Policy 1992 set the 1.6 gpf toilet (6 litres) as the minimum requirement for installation in new residential buildings (U.S. EPA, 2008). Today, manufacturers have reduced the flushing volume to as low as 1.1 gpf (4.2 litres), without sacrificing basic requirements such as complete waste removal and drain line carry (Williams *et al.*, 2011).

Instead of using 1.6 gpf, high efficiency toilets (HETs) allow a further 20% savings by using 1.28 gpf (4.85 litres). HETs are available in different flush types, including the conventional single-flush gravity fed, single-flush pressure assist, and dual-flush in both gravity fed and pressure assist. Designed for light and heavy flushes, dual-flush toilets usually consist of two buttons or levers, one for low flush (3 litres) for liquid waste and the other for high flush (6 litres) for solid waste. Air assisted toilets make use of the pressure of compressed air to remove waste with minimal water as low as 0.5 gpf. Although such toilets are used in aircrafts, use of air-assisted toilets in homes is less widespread because of the need for air lines, a compressor, and the higher initial costs of air-assisted units. Increased education and marketing efforts may result in wider adoption of these highly efficient toilets (Sharpe & Swistock, 2008).

Other types of toilets, which are not in as widespread use, may also significantly reduce the use of water. Composting toilets do not use any water, while foam-flush toilets, a variation, use a small amount of water. Urine-diverting toilets do not use water for liquid waste but require behaviour change (Schlunke *et al.*, 2008). Another water-saving feature is an integral hand basin that allows a person to wash his or her hands with water that flows into the tank and that then can be used for the next flush (Figure 5.2). The Australia and New Zealand water labelling programs give five stars to toilets only if they have these integral hand basins (Standards Australia/ Standards New Zealand, 2005). It is important to note that the use of these products has implications for drain line carry issues as they reduce other flows in the pipe.

Figure 5.2 Caroma water saving toilet with integrated hand basin (Caroma, 2013).

5.1.2 Shower heads

Currently, there are four main types of showerheads: (1) bath and shower mixers, in which the hose and spray are attached to your bath and the temperature and amount of water are adjusted through the taps; (2) manual mixer showers, where the hose and spray come out of a wall unit and there's a temperature control that mixes the hot and cold water supply; (3) electric shower, which rapidly heats cold water as it flows towards the shower head; and (4) power showers, which use an electric pump to adjust the pressure and water temperature (Which?, 2013).

The UK Sustainable Shower Study by Unilever UK and Ireland recorded accurate water consumption data for 2600 showers in 100 British households. The study found that while an average 8 minute shower uses about 62 litres of water, an average bath uses about 80 litres of water. However, some power showers can use up to 136 litres of water in the same time, busting the myth the showers are always more water efficient than baths (Unilever, 2011).

A low flow showerhead typically uses about 8 litres of water per minute. A showerhead labelled by WaterSense in the United States has a maximum flow rate of 2 gpm (7.6 litres/minute) at 80 psi (U.S. EPA, 2010). Water efficient showerheads usually conserve water by one of the two following ways. Commonly, the water is aerated by infusing air into the water flow. This reduces water consumption by up to 50% by replacing some of the water with air bubbles. The main performance issue for these showerheads is their ability to provide enough pressure for rinsing or other tasks. The inability to rinse in a reasonable amount of time possibly led to increased showering time, potentially offsetting some gains from conservation, as well as removal of the showerheads, which would eliminate all gains from conservation. Due to the increased air flow introduced into the water stream to reduce flow rate, users often complain of getting cold spots and as the water is forced through minute pores, users may feel the water like piercing needles. The second type of water efficient shower heads use a pulsating technology, with which the water is supplied at a high pressure but instead of a continuous flow, it is turned off and on 30 times per second. The shower experience feels normal because the pulsating is fast enough for interruption in water flow to go unnoticed. Pulsating shower heads are more expensive than aerating shower heads (How to Save Water, 2013).

While water-saving devices can typically reduce end use water consumption, there is evidence that human behaviour can diminish engineered savings. Inman and Jeffrey (2006) found that the installation of water-saving devices led to an increase in water consumption due to the residents' misguided belief that they were saving water through their efficient devices and hence, took longer showers. Stewart *et al.* (2013) studied households' response to an alarming visual display device, which provided a digital readout of shower parameters such as flow rate, duration and temperature. The device also presented an illustration of the volume of water remaining, before the alarm will sound, enabling users to monitor the progress of their shower against a set target. Compared with written feedback such as quarterly bills, electronic devices provide quicker and more frequent feedback, thus better informing the consumer of the consequence of their specific behaviours (Darby, 2006; Midden *et al.*, 2007). By using smart meters to record shower end use event characteristics (e.g., volume, duration and flow rate) before and after the intervention in 44 households in the Gold Coast Residential End Use Study, the research found that shortly after the intervention, there was a mean reduction of 15.40 litres (27%) in showering volumes. However, two subsequent smart metering reads indicated that shower end use water consumption savings diminished over time and mean showering volumes reverted back to their pre-intervention level after 4 months. That is, the longitudinal study provides empirical evidence that technological devices informing resource consumption may not be effective unless instilled habits or attitudes can be also modified (Stewart *et al.*, 2013).

5.1.3 Faucets

A WaterSense labelled bathroom sink faucet use a maximum of 1.5 gpm (5.6 litres) compared to the standard flow of 2.2 gpm (8.3 litres) (U.S. EPA, 2007). In order to avoid consumer inconvenience arising from very low flow, WaterSense has specified a minimum flow rate of 0.8 gpm (3 litres/minute) at a pressure of 20 psi. Compared to conventional taps, low flow faucets have a flow restrictor inside with a spinner attached which aerates the water flow so that users think they are getting more water than they are. Self-closing taps or tap sensors are useful for public buildings where taps are carelessly left on. Some studies show that the inclusion of a sensor increases water use (Hills *et al.*, 2002; Gauley & Koeller, 2010). Hills *et al.* (2002) note that conventional swivel-top taps used less water than both push-tops and infrared taps, although push top use was reduced after switching from a 15 to a 7 second delay (Williams *et al.*, 2011).

5.1.4 Washing machines

The water and energy efficiency of washing machines have significantly improved over the last two decades. The water efficiency of washing machines depends on the manufacturer or model of the product as well as the program setting used while washing at home. Today, while an average washing machine uses about 10 litres of water per kg of clothes at the standard 40°C cotton setting, the water consumption of washing machines show wide variations and usually ranges from 6 litres per kg of clothes to 14 litres per kg (Which?, 2013). Traditionally, front loading washing machines were more water and energy efficient than top loading ones. Compared to a top-loading machine which requires enough water to cover all the clothes in its drum, a front-loading washer with a horizontal drum needs only a third of that amount. As the drum turns, it uses gravity to drop the clothes back into the water. Moreover, while a top-loading machine empties the soapy water and refills for a rinse agitation cycle, a front-loading machine sprays clean water on the load as the drum continues to turn (Heste, 2012). Apart from the water efficiency aspect, front loading washing machines save space by allowing dryers or other things to be mounted on top and are considered to be gentle on clothes. However, top loading machines are comparatively cheaper, ergonomically convenient and have the option of adding extra clothes in the middle of a washing cycle.

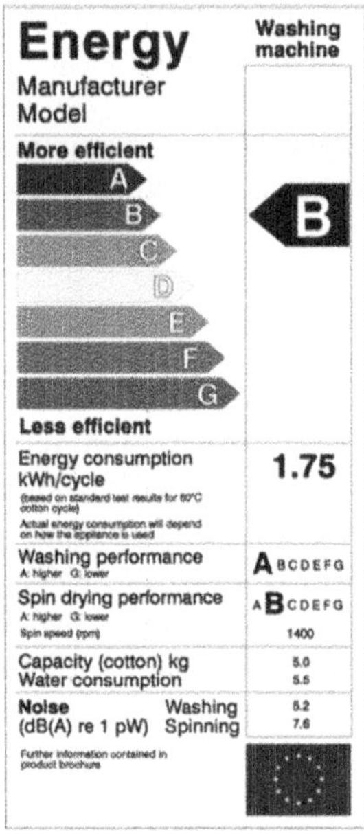

Figure 5.3 Sample EU energy label for a washing machine.

Nowadays, most big brands provide information on the water efficiency level of the washing machine, either through water or energy efficiency labels displayed on the product or through product specifications on the website or user manuals. The EU energy efficiency label, launched in 1995 and revised in 2010, rates products from A (most efficient) to G (least efficient) based on the total energy consumption (Figure 5.3). Although the label contains other relevant features, such as water consumption and noise levels for washing machines, the rating is only based on energy efficiency (DEFRA, 2010). Similarly, the energy star label is a U.S. Environmental Protection Agency voluntary program that helps businesses and individuals save money and protect the environment through superior energy efficiency. The energy star criteria for washing machines, originally launched in May 1997, have been amended in February 2013. It currently requires residential clothes washers to have a Modified Energy Factor of ≥2.0 and a Water Factor of ≤6.0 (where the water factor is the quotient of the total weighted per-cycle water consumption, divided by the capacity of the clothes washer) (Energy Star, 2013). Clothes washers that have earned the ENERGY STAR are about 20% more efficient than non-qualified models and are more efficient than models that simply meet the federal minimum standard for energy efficiency. Sammer and Wüstenhagen (2006) used a discrete choice approach and surveyed a total of 151 consumers in Switzerland to investigate the relative importance of various product attributes in consumers' decisions while purchasing washing machines. In this study, while 31.8% of respondents mentioned price and 11.9% mentioned energy consumption as their first priority, only 7.9% considered water efficiency as the most important criteria in their decision making.

5.1.5 Urinals

Urinals are mainly used in public toilets and are rarely installed in a residential dwelling. With traditional urinals, flush valve operation can be improved to provide water savings. Australian technical standards prevent the use of automatic or set-cycle flushing and require that for flush on-demand operation the motion sensor prevents interference (Standards Australia, 2005). The Australia and New Zealand water efficient label requires smart demand operation to get more than three stars, and a urine-sensing device to get six stars (Standards Australia/Standards New Zealand, 2005). Use of sensors for urinals may actually increase water use if the sensors are not installed correctly or do not operate properly (Hills *et al.*, 2002; Gauley & Koeller, 2010). A urine-sensing device may be less prone to ghost flushing than sensors that detect humans, but no studies on this subject were found.

High efficiency urinals use as low as 0.5 gallons per flush. Waterless urinals use a special drain insert that traps urine below a blue liquid and forms a barrier against sewer vapour escape. Urine sinks below the blue liquid because it is heavier. However, these urinals have experienced some problems with crystallisation of urine. The U.S. industry standard provides for testing for resistance to stoppage from debris and ammonia detection in the air near the waterless urinal, but does not address crystallisation (American Society of Mechanical Engineers, 2006). Australia and New Zealand do address crystallisation by requiring a minimum of two additional water-using fixture units upstream of every waterless urinal (Cummings, 2009). Another possible solution to alleviate the crystallisation problem is to use a very small amount of water, such as one cup, with these urinals.

5.1.6 Dishwashers

Like washing machines, dishwasher models also differ greatly in terms of their water and energy consumption. Richter (2010) reported that between 1990 and 2005 the water usage of dishwashers more than halved and between 1999 and 2005 the average energy consumption per dishwashing cycle decreased by about 30%. According to Which? (2013), which tests and compares appliances, the water usage of dishwashers, under a normal setting, ranges from 10 to 20.9 litres per cycle. Using the eco or energy saving option reduces the water usage to a range of 7 to 17 litres per cycle (Which?, 2013).

Washing manually under a running tap or with water stored in the kitchen sink can be quick and easy for cleaning a few dishes or utensils. However, a fully loaded dishwasher can be more water efficient compared to washing the same number of dishes by hand. Berkholz *et al.* (2010) studied the washing up behaviour of 150 participants from all regions of UK. The authors found that while manual washing by hand used 49 litres of water and 1.7 kWh of energy on average, the dishwasher used 13 litres of water and 1.3 kWh of energy on average for the same amount of dishes under the conditions tested. Moreover, manual washing required more time (about 60 minutes) compared to the time taken for loading and unloading the dishwasher (9 minutes) and also resulted in relatively less clean dishes. In a subsequent study, Berkholz *et al.* (2013) involved 289 participants from 29 countries to investigate the individual attitudes towards dishwashing. Among all the participants, one-third had a dishwasher at home; their main reasons for purchase included the time-saving (69%), the need to wash a large number of dishes (45%) and the better cleaning result (36%). Only less than one-third of the dishwasher owners mentioned energy and water conservation as one of their primary motivations. However, the water saved from dishwashers is often offset when users pre-rinse their dishes before loading the machine. Richter (2010) found that in Germany and the UK majority of the users scraped off the food leftovers with kitchen paper, whereas in Sweden and Italy, more users quickly pre-rinsed each item before placing it to the dishwasher.

BOX 5.1 WATER CONSERVATION USING WATER EFFICIENT DEVICES IN THE UNITED STATES

The Residential End Uses of Water Study (REUWS), conducted by the American Water Works Association Research Foundation (AWWARF) in 1999, was designed to obtain specific water end use data from a sample of 1188 single family households across 12 locations in North America. The study calculated the mean daily fixture usage for toilets, showers, clothes washers, dishwashers, baths and faucets. According to the data from all 12 sites, the average toilet flushing rate was 5.05 times per person per day and the combined bath and shower rate was 0.75 per person per day. Clothes washers were run an average of 0.37 times per person per day and dishwashers were run an average of 0.1 times per person per day. Faucet utilisation was about 8.1 minutes per capita per day and the average loads of laundry per day was 0.96.

However, the volume of water used for each of the above activities directly depended on the water efficiency of the fixtures in use. For instance, the average volume per load of clothes was 40.9 gallons (154 litres) with a standard deviation of 12.2 gallons and a median volume of 39.8 gallons (150 litres). About 14.5% of the flushes were

(*Continued*)

BOX 5.1 (*Continued*)

less than 2.0 gpf (7.6 litres), 34.7% of the flushes were between 2 and 3.5 gpf (7.6–13.2 litres), and 50.8% were greater than 4 gpf (15.1 litres). The low flow shower homes (15% exclusively) used an average of 8.8 gpcd (33.2 lpcd) for showering, while the non-low flow (24.5% exclusively) shower homes used an average 13.3 gpcd (50.2 lpcd). However, the duration of the average shower in the low flow shower homes was 1 minute and 48 seconds longer than the average shower duration in the non-low flow homes. The Table 5.1 shows an estimate of the volume of water that a person can save annually by replacing conventional appliances with water efficient fixtures described above.

Table 5.1 Estimated water savings from various water-saving devices.

Fixture	Frequency of use per person	Daily water use without water-saving fixture (gallons/ person)*	Daily water use with water-saving fixture (gallons/ person)	Daily water savings with water conserving devices (gallons/ person)	Annual water savings (gallons/ person)
Low Flush Toilet (1.6 gpf)	5.1 flushes/ day	20.4	8.2	12.2	4453
Low Flow Showerheads (2.5 gpm)	5.3 mins/day	15.9	13.3	2.6	949
Low-flow faucets (1.5 gpm)	4 mins/day	12	6	6	2190
Front loading washing machine (27 gpl)	0.37 loads/ day	18.9	10	8.9	3249
Water Efficient Dish washer (7 gpl)	0.1 loads/day	1.1	0.7	0.4	146
Total		68.3	38.2	30.1	10,987

*Assumes conventional toilets at 4 gpf, showerheads at 3 gpm, washing machine at 51 gpl, and dishwasher at 11 gpl.

Source: Adopted from Vickers (2001).

5.2 ADOPTION OF WATER EFFICIENT DEVICES

In order to promote the adoption of water efficient devices, many governments and water agencies have conducted public awareness programs, launched labelling of water appliances with their consumption specifications and also given rebates for the installation of these devices. Installation of water-efficient devices is seen as an effective manner of inducing water conservation for several reasons. First, water consumed through both indoor and outdoor appliances (e.g., showers, toilets, washing machine, sprinklers) represents a significant share of households' daily water use. Second, the reduction potential of water saving fixtures is now well acknowledged; a water-efficient washing machine may use only one-third the water of an inefficient model, a dual flush toilet uses a quarter of water on half flush compared to an old-style single-flush toilet that requires 12 litres of water per flush, and a water-efficient shower head can use only 7 litres per minute compared to a standard shower head that uses up to 25 litres of water per minute. Third, policies to promote installation of water efficient devices are likely to be more politically acceptable than price increases or policies imposing water restrictions. Finally, another reason why adoption of water efficient equipment is a potentially interesting policy tool is the pervasive role of habits in human behaviour which may make other forms of non-price policies, such as public information campaigns, yield little effect (Thøgersen & Ölander, 2002).

A number of studies have focused on the factors that influence water conservation behaviours as well as the barriers to adoption of water saving devices. In a July 2005 study of factors that influence water conservation in Melbourne, Clarke and Brown (2006) surveyed 2600 residents of the Bayside area. They focused on three water use behaviours/appliances, finding that 6% of respondents had a rainwater tank, 52% had a water efficient showerhead, and 5% reuse greywater. With regard to water behavioural change they found that this is dependent upon the capacity of individuals to acquire and apply water saving and reuse measures. The authors report that the key barriers to a widespread practice of conservation and alternative water use include difficulty of implementation, cost, and dwelling ownership or rental status, rather than issues of attitudes and lack of awareness. Using results from a survey study of 1495 people, Dolnicar and Hurlimann (2010) showed that although Australians have positive attitudes towards water conservation and water-saving appliances, these attitudes do not always result in actual behaviour. The perception of inconvenience and impracticality and the costs associated with purchasing water-saving equipment were identified as the main barriers for the adoption of water saving behaviours.

Hypothetically, there are two counter arguments regarding the effect of household income on adoption of water saving devices. While richer households can afford to purchase water-saving devices, due to their relatively higher incomes they have a lower incentive to conserve water and a higher opportunity cost in terms of the time invested to retrofit their water appliances (Berk *et al.*, 1993; Millock & Nauges, 2010). Lam (2006) conducted surveys on intentions to adopt a dual-flush toilet in Taiwan in 2002 (166 residents) and in 2004 on a different sample (210 residents) and found that high income had a significant positive effect on intentions to adopt the equipment in one sample, and was insignificant in the other sample. Lam (2006) found that people's intention to install a dual-flush toilet was related to a strong perception of environmental threat, a strong belief in the efficacy of adopting a dual-flush controller compared to alternative strategies, their estimation of the number of other residents that would take action to save water and their personal efficacy to save money by conserving water.

The effects of education on the adoption of water efficient technologies are quite ambiguous. According to Berk *et al.* (1993) household members must understand the nature of water shortages and water-saving options, and believe that there personal efforts will bring about some water conservation and will not be offset/exploited by others. As education is likely to be linked to these characteristics, better education households are expected to conserve more. While De Oliver (1999) found education to be positively related to water conservation, Lam (2006) showed that education had no effect on the behavioural intention to conserve water.

Using survey data of approximately 10,000 households from ten OECD countries, Millock and Nauges (2010) assessed the impact of the water charging system, that is, the tariff structure and rate, and the impact of applicable environmental labels. In general, households displayed bad knowledge of their water bill, which suggests that the price of water as such is not sufficient to explain the adoption of water-efficient equipment. On the other hand, the authors found a clear-cut result of the structure of water charging in terms of metering or not. Households that were both metered and charged for their water individually (volumetric fee) have a much higher probability of investing in indoor water-efficient equipment compared to households that are not charged for their water, or, to a smaller extent, compared to households that are charged but not metered individually for their water (flat fee).

The study by Renwick and Archibald (1998) in two communities in California also showed a strong and significant positive effect of the marginal price for water on the number of water-efficient equipment in the household. However, Syme *et al.* (2000) concluded that as long as water is under-priced, monetary savings are not a big factor in water conservation. Subjective norms or other socially motivated values are more important in strengthening behavioural intentions to conserve water. However, an actual or expected price increase could foster investment in water-efficient devices since it has a direct impact on the water bill.

In general, owners of dwellings are more likely to invest in water efficient technology that tenants. Firstly, owners have the right to make permanent changes in fittings within the household while tenants are more constraint. Secondly, owners can benefit from the long-term reduction in water bills and also enjoy the price premium while re-selling the house. Some authors have found a positive and significant relationship between ownership and the adoption of water-efficient equipment (Gilg & Barr, 2006; Millock & Nauges, 2010). In the 2004 study in Taiwan, Lam (2006) found that residents in detached houses had less incentive to conserve water than those living in apartments, as the former did not have to share their water tanks with neighbours. Additionally, Millock and Nauges (2010) found that residents of older buildings are less likely to adopt water efficient devices.

Martínez-Espiñeira and García-Valiñas (2012) studied the determinants of the adoption of water-efficient devices and of water-saving habits using data from 27,000 households in Spain. The authors found that households with higher levels of education and income, and those with full-time workers and large families tend to adopt a larger number of water-saving devices. However, the presence of elderly people in the house diminishes the chances of adopting these technologies. Their findings suggest that public policies aimed at promoting the adoption of pro-environmental habits and investments associated with water use might be particularly effective if focused on older males with low educational levels. In contrast to other studies that found women to be more pro-environmental, Lam (2006) found that gender had no effect on water conservation behaviour. The authors concluded that this is probably related to the behaviour in question, as retrofitting toilets may be considered more masculine than other environmentally friendly and healthy consumption habits.

Thus, with respect to socio-demographic variables, the evidence in the literature has been quite ambiguous. Other than the ownership of dwellings, there is paucity of strong evidence on the relationship between income, education, price, gender or age and the adoption of water conservation behaviour through use of water efficient devices.

BOX 5.2 EFFECT OF WATER EFFICIENT APPLIANCES ON DIURNAL DEMAND PATTERNS IN QUEENSLAND, AUSTRALIA

Knowledge of the typical water consumption volumes and the associated end uses (e.g., toilet, tap, shower, clothes washer, irrigation and leaks) allow a more proactive approach to demand management that can be used for targeted water conservation strategies and schemes (Lahlou & Attia, 2005; Beal *et al.*, 2011; Willis *et al.*, 2011). End use data is also required to accurately determine how water is consumed by residents within their home and is needed to continue to improve the development of demand forecasting and water distribution network models (Jacobs, 2008; Blokker *et al.*, 2009). Water demand modelling and forecasting has a range of purposes, which include: planning new (greenfield) developments or water distribution network expansion; estimating the capacity and operation of water storage reservoirs; pump and pipe sizing; as well as aiding decision making when addressing metropolitan water management challenges for example, pricing policy or water use restrictions (Bougadis *et al.*, 2005; Blokker *et al.*, 2009).

Residential water use does not remain constant, but experiences periods of high and low demand throughout the 24 hour period. In order to investigate the impacts of a household's stock of water efficient devices on the diurnal water demand patterns, the South East Queensland Residential End Use Study (SEQREUS) used high resolution smart water meters and remote data logging equipment in a sample of 191 households from Gold Coast, Brisbane, Ipswich and Sunshine Coast in Queensland. These meters can provide up to 0.014 litres per pulse data output recorded at 5 s intervals which can then be stored by the attached data logger and sent for offsite analysis via general packet radio service (GPRS) (Beal *et al.*, 2010). In addition to the water meters, household water audits and water use diaries are also conducted, in what is termed a mixed method approach, to allow for the development of household

(*Continued*)

BOX 5.2 (*Continued*)

water stock inventories and accurate characterisation of residential water end uses (Willis *et al.*, 2009; Beal *et al.*, 2011).

During the household water audits, the efficiency specifications of the household's stock of water appliances were recorded based on the information displayed on the WELS labelling of the appliances. The aggregate efficiency scores of the households were calculated and the households were then clustered into two groups – those with efficiency ratings ≥3 stars and those with ratings <3 stars. The study found that the average daily household and per capita water consumption for the 191 participating households was 333.7 litres per household per day and 132.61 litres per person per day (Carragher *et al.*, 2012).

The <3 star efficiency cluster diurnal demand pattern showed a peak morning demand between 8–9 am of 12.37 litres per person per hour (l/p/h). The bulk of water consumption during the morning peak hour was attributed to shower and clothes washer which, when added together, contributed to over 65% of the total demand. The <3 star households experienced a less dramatic evening peak demand between 6–7 pm of 9.31 l/p/h. The primary contributor to this evening peak demand was shower consumption which contributed to over 43% of total water use in that particular peak hour. Tap, toilet and clothes washer use all exhibited fairly constant use over the course of the afternoon and into the evening (3–7 pm) (Carragher *et al.*, 2012).

The ≥3 star cluster showed similar trends, although there was a slightly later and less dramatic morning peak demand (9–10 am) of 10.47 l/p/h, when compared with the <3 star cluster. The bulk of the water consumption during the morning peak period was again primarily composed of shower and washing machine use, however, greater tap and toilet consumption occurred which contributed to over 27% of the peak hour demand. The evening peak occurred between 6–7 pm with a demand value of 7.76 l/p/h which had a noticeable reduction in shower demand in this peak hour, when compared to <3 star cluster (Carragher *et al.*, 2012).

Comparison of the diurnal demand patterns for the ≥ and <3 star efficiency household groupings show that water demand for the less efficient households is greater than that of the more efficient households for most hours of the day. The reduction in the peak water consumption due to the household composite star rating for the sample being ≥3 stars was 15.35% for the morning and 16.64% for the afternoon (Carragher *et al.*, 2012). Reductions in peak demand have substantial implications to water distribution networks. There is the potential for significant capital efficiency opportunities as a result of installing smaller diameter pipe infrastructure in new developments as well as potential deferment of existing supply network's inevitable future upgrade costs.

5.3 WATER EFFICIENCY LABELLING

Around the mid-2000s, certain developed countries, such as the United States, Australia and Singapore, introduced water efficiency labelling schemes to allow consumers to make informed choices about the water appliances they purchase. These labels specify the water consumption or flow rates of the fixtures and also highlight their efficiency levels through tick marks or stars. Many of these programs started as voluntary and were made mandatory over the years. The range of products covered under the scheme also increased. The WaterSense labelling program in the United States has been discussed in Section 4.3. In this chapter, Box 5.3 and 5.4 provides an overview of the Water Efficiency Labelling Scheme (WELS) in Singapore and Australia respectively.

BOX 5.3 WATER EFFICIENCY LABELLING SCHEME IN SINGAPORE

In Singapore, the Water Efficiency Labelling Scheme (WELS) is part of an ongoing effort by the Public Utilities Board (PUB) to encourage people to use water wisely and make water conservation a way of life. WELS enables consumers to make informed purchases based on the water efficiency of a product. By spurring consumers to choose higher rated water efficient products, WELS also gives manufacturers and suppliers an incentive to introduce more water efficient products into the market.

WELS was first introduced as a voluntary scheme in 2006. As a follow-up to the voluntary scheme, the mandatory scheme was implemented in 2009 (Figure 5.4). Since 1 July 2009, all basin, sink and shower taps, dual flush low capacity flushing cisterns and urinals are rated as either 'GOOD', 'VERY GOOD' and 'EXCELLENT' under the WELS (Table 5.2). The number of ticks is used to distinguish between the ratings whereby a label with more ticks indicates a more water-efficient product. With effect from 1 Oct 2011, clothes washing machines have also been moved over to the Mandatory WELS from the Voluntary WELS (Public Utilities Board (PUB), 2011a). Sales figures from a market survey by GFK Asia Pte Ltd. shows that 74% of the washing machine models available in the market are water efficient (i.e., they have at least one tick). From April 2014, it is mandatory to sell washing machines with at least one tick. Only showerheads currently remain under the Voluntary WELS.

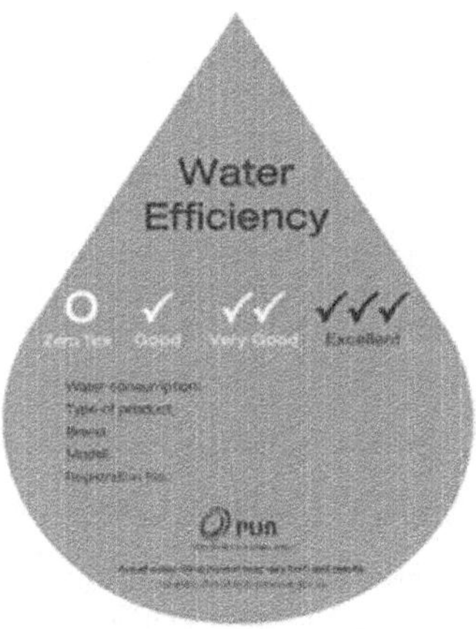

Figure 5.4 Singapore water efficiency labelling.

(Continued)

BOX 5.3 (*Continued*)

Table 5.2 Water efficiency labelling scheme in Singapore.

Fittings	Zero Tick O	Good √	Very Good √√	Excellent √√√
Shower taps, mixers and showerheads (l/min)	>9	>7–9	>5–7	5 or less
Basin taps and mixtures (l/min)	>6	>4–6	>2–4	2 or less
Sink/Bib taps (l/min)	>8	>6–8	>4–6	4 or less
Flushing cisterns – dual flush types (l/flush)	N/A	>4–4.5 (full flush) >2.5–3 (low flush)	>3.5–4 (full flush) >2.5–3 (low flush)	>3.5 or less (full flush) >2.5 or less (low flush)
Urinals (l/flush)	N/A	>1–1.5	>0.5–1	0.5 or less or waterless urinals

Source: Public Utilities Board (PUB), (2011c).

BOX 5.4 COSTS AND BENEFITS OF WATER EFFICIENCY LABELLING SCHEME IN AUSTRALIA

As one of the key programs to address water shortage and droughts in Australia, government agencies and water utilities introduced the Water Efficiency Labelling and Standards Scheme (WELS) in 2006 (Figure 5.5). It is administered by the Department of the Environment, Water, Heritage and the Arts (DEWHA), in association with the State and Territory governments. According to the scheme, manufacturers of toilets, washing machines, dishwashers, urinals, faucets and showers are required to display a star rating of the water efficiency of these products.

Even before WELS was launched, manufacturers and importers had to comply with different standards and regulations. For example, the WaterMark scheme is a mandatory certification which ensures that plumbing and drainage products are fit for purpose and appropriately authorised for installation. Since 1993, AS 3500 has specified the maximum water use per flush for toilets and since 1998, energy labelling has been mandatory for washing machines and dishwashers.

WELS allows consumers to differentiate between products on the basis of their water efficiency and enables them to make informed purchase decisions. WELS also acts as a basis for rebate programs, by determining the extent of rebates offered and the uptake rates. However, this demand management tool has several administrative and indirect costs associated with it. Table 5.3 outlines the costs and benefits to various stakeholders involved in the process.

(*Continued*)

BOX 5.4 (*Continued*)

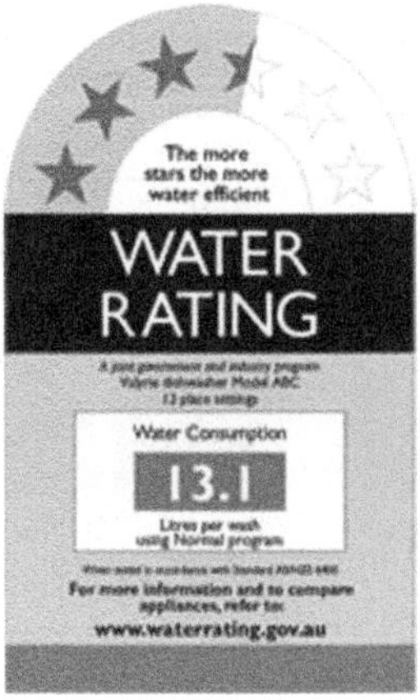

Figure 5.5 Sample water efficiency label in Australia.

Table 5.3 Costs and benefits of WELS in Australia.

Stakeholder	Cost or benefit	Description
DEWHA	WELS administration costs	Costs involved in promotion, enforcement activities, database administration, staffing and overheads.
	WELS registration fees	AUD1500 per model or family of models registered.
Suppliers	Other registration costs	Staffing requirements to undertake registration process, including preparing documentation.
	Additional labelling costs	Printing and affixing labels.
	Additional testing costs	Accredited testing for water-efficiency rating.
	Change in cost of products sold from suppliers to retailers.	Due to any price premium for higher water-efficiency products.
Retailers	Additional retailers' costs	Staffing requirements to check labels, train staff. Transition between unlabelled and labelled stock.

(*Continued*)

BOX 5.4 (*Continued*)

Table 5.3 Costs and benefits of WELS in Australia. (*Continued*)

End-use customers	Change in product cost sold from retailers to end-users (possibly through installers)	Due to any price premium for higher water-efficiency products.
	Change in energy bills	Due to reduced water heating required (energy bills).
Water utilities	Change in water bills/ revenue	Due to reduced water consumption (water bills).
	Change in operating costs	Due to reduced pumping for water supply and sewage treatment.
Global	Reduced greenhouse gas impact	Due to reduced energy use associated with hot water use.

Source: Chong *et al.* (2008).

On behalf of DEHWA, the Institute of Sustainable Futures (Chong *et al.*, 2008), conducted a cost-benefit analysis of WELS, in order to analyse its effectiveness as an urban demand management tool. The findings of the study are discussed below.

Effect on water consumption

In response to the millennium drought that affected most regions in Australia, WELS was introduced along with several other demand management policies to curb water consumption. Hence, it has been difficult to elicit the individual contribution of WELS in reducing demand. However, in general, the mandatory WELS program have encouraged consumers to purchase water efficient products and also enable nationwide implementation of certain building regulations and minimum standards targeting water efficiency.

Compared to a baseline of no WELS and only voluntary labelling, it is estimated that WELS will reduce wide water consumption by 800 million m^3 in the whole nation between 2005–06 and 2020–21. When disaggregated by fixtures, the showerheads (36%) and washing machines (35%) have the greatest potential for conservation. This is followed by toilets and urinal (23%) and faucets (6%). The difference in relative contributions is due to the amount of water used by different products.

Costs to administrators and suppliers

The largest share of the direct costs associated with WELS are to be borne by the administrators of the scheme and the suppliers of WELS products. Between 2005–06 and 2020–21, DEHWA, the WELS regulator, is estimated to incur a cost of AUD 16 million for staffing, promotion, enforcement, and database management. Labelling

(*Continued*)

BOX 5.4 (*Continued*)

costs (around AUD 7 million) and registration fees (AUD 5 million) will form the bulk of the total costs to suppliers estimated at around AUD 16 million. However, these estimated are based on a number of assumptions about future activities, which are uncertain. There is also uncertainty regarding the existing and potential price premiums due to WELS. However, price premiums are not likely to be substantial and long-lived because, at each star rating, various products are available at different prices. Moreover, an increase in demand will also increase supply, pulling down the market price further.

Sustainability and financial benefits

While consumers of WELS products are likely to gain from reduced water and related energy bills, they will also contribute to reducing greenhouse gas emissions. Between 2005–06 and 2020–21, WELS is estimated to result in 9 million MWh of energy savings and 6 million tonnes of avoided greenhouse gas emissions.

Cost-effectiveness

Compared to other water management options, which have been planned or implemented in various regions in Australia, WELS is cost-effective. The unit cost of WELS is estimated to be between AUD 0.08–0.21/m^3, depending on the discount rates and supplier costs. On the other hand, the costs of desalination ranges from AUD 1.19 – AUD 2.55/m^3, and wastewater recycling costs about AUD 5.50/m^3.

Source: Chong *et al.* (2008)

5.4 REBATES TO PROMOTE WATER EFFICIENT FIXTURES

The use of rebate programs is a popular demand management approach undertaken by energy and water utilities. Rebates programs subsidise the adoption of energy or water efficient devices by refunding the cost of the appliance partially or fully. This non-price tool is used to encourage residential consumers to purchase water saving devices when they are installing or retrofitting these appliances. For example, the U.S. Environmental Protection Agency's (EPA) website contains a database of all the current rebate programs offered by different utilities across all states in USA (U.S. EPA, 2013). Although some utilities, such as the Los Angeles Department of Water and Power (LADWP) (see Box 4.1), claim that rebates are a cost effective means of water conservation compared to other supply side alternatives such as desalination, wastewater recycling and importing water, the true effectiveness of such policies is questionable at times. This is because the desired voluntary activity, which, in this case, is the installation of water efficient devices, may have also taken place in absence of the rebate policy.

In their study involving a water utility in North Carolina, Bennear *et al.* (2013) aim to identify the 'additionality' of a high efficiency toilet (HET) rebate program.

Here 'additionality' refers the degree to which the rebate program has resulted in action that would have occurred otherwise as well. The authors used a database of 683 households that participated in the rebate program; the database contained information of the households' water consumption over a period of 3.5 years after the retrofit as well as a survey of the households' motivation for replacing their previous toilets. The study found that only 37% of the water conservation associated with the HET can be directly attributed to the rebate program, while the rest would have taken place anyways because the households were already planning to retrofit their toilets even in absence of the rebates. Moreover, significant water savings of upto 7% were only achieved when a pre-1991 toilet was replaced by a HET. The authors estimated that while new capacity expansion would have cost USD 7 per 1000 gallons of water, the rebate program cost about USD 11–15 per 1000 gallons. Hence, from a cost effectiveness perspective, this policy proves to be a poor demand management tool (Bennear *et al.*, 2013).

5.5 CONCLUSION

Technological solutions to water conservation in urban households started to gain popularity since the 1990s, when newly engineered water fixtures and appliances promised to save significant amounts of water compared to their previous counterparts. In order to promote the installation of such fixtures, government agencies and water utilities have formulated various legislations and standards, building codes, water efficiency labelling schemes and rebate programs. During the initial period, the users of such water efficient products faced certain problems, such as requiring to flush more than once to remove waste; however, today advancement in technology have dealt with these issues, providing consumer satisfaction as well as enhanced performance. Now, most products available in the market are water efficient and there are less price premiums for purchasing products with greater water efficient. The penetration of water efficient products have greatly increased in recent years and manufacturers have almost reached the minimum amount of water consumption for most appliances. Now, technological solution to urban water management is heading towards greater recycling of wastewater, either centrally by water agencies or de-centrally at household level.

Chapter 6

Demand side management tools: Comparing price and non-price approaches

Simple water conservation practices, such as keeping the tap shut while brushing or using the washing machine with a full load of clothes only, are often advertised on water agency websites, on television or newspapers and communicated to consumers through leaflets accompanying the water bills. However, altering water use behaviour within the household, involves implementation of several demand management tools that address the basic factors that determine water consumption patterns of end users. Water conservation policies have been in place in several cities as early as the 1960s and through research and experience, these policies have evolved over time to become more specific to the local challenges. The bundle of price and non-price water conservation policies can be generalised into four main categories:

(1) *Price* – Changing the pricing structure or rate provides economic incentives to reduce wastage of water within the household. Conservation oriented pricing usually involves increasing block tariffs, conservation taxes or seasonal water pricing that penalises consumers for using higher volumes of water during normal times or during periods of water shortage. However, for price changes to be effective in reducing demand, water consumption should be individually metered and information about the rates and usage should be communicated with consumers through the water bills. The responsive of demand to changes in price depends on a wide range of factors, such as household income, initial level of prices, dwelling types and proportion of indoor and outdoor uses.

(2) *Regulation* – Regulatory policies encompass voluntary and mandatory water use restrictions, for specific activities during particular times of the day. These restrictions are usually implemented during periods of acute water shortage and the level of restrictions enforced depend on the severity

of the water crisis. Restrictions usually target outdoor water uses, such as lawn watering through sprinkler or hoses, car washing and filling up swimming pools. In order for restrictions to be effective, the prevailing restrictions need to be widely communicated to the public through the media and also strictly enforced by law enforcing agencies. Other forms of regulations include state or local level ordinances that mandate the sale and installation of water fixtures meeting specific low flow requirements.

(3) *Technology* – Engineering approaches to reduce water consumption refer to installation of water efficient fixtures, such as low-flow faucets, showerheads, toilets and washing machines, which consume less water for achieving the desired task. While retrofitting whole appliances can be costly, installing water conservation kits, such as flow aerators in faucets and containers within toilet cisterns, can be temporary solutions. Promotion of water efficient devices also require communication with consumers and this is usually done by labelling water appliances and by providing rebates for installing them. Water efficiency labels contain information about the appliance's water consumption rate and relative efficiency. Besides enabling consumers to make informed purchase decisions, these have become environmental marketing tools for manufacturers.

(4) *Information* – Information campaigns are designed to educate water providers, adult consumers and school children about the importance of water as a precious scarce resource and the need to conserve it. Information aims to change attitudes and behaviour towards water consumption and inculcate a strong conservation culture within the society. The delivery of information can take several forms: school curricula and competitions can disseminate information to children; religious preaching, advertisements and observation of world water or environment days can raise public awareness; and workshops or trainings can educate water managers or community leaders.

It is often difficult to generalise the effectiveness of price and non-price based approaches in promoting water conservation and to conclude with certainty the success of one over another. There have been substantial debates among academics and policy makers about the effectiveness of these policy instruments, especially with respect to efficiency and equity among residential consumers. A wide range of research studies have tried to estimate the outcomes of individual demand management policies and have resulted in various numerical approximations of the relative effectiveness of each of the policies. While the lack of well documented data provides a basic obstacle for many researchers, the implementation of several policies within the same time period makes it challenging to elicit the contribution of the each individual element.

The seasonal water pricing and the water conservation policies implemented in Los Angeles have been discussed in Box 3.5 and 4.1, respectively. Hanemann and

Nauges (2005) used a dataset covering water consumption of all residential users in Los Angeles during the drought (1988–1992) to estimate households' responsiveness to various policy tools. The authors found that while households were generally responsive to the conservation measures, the magnitude of the responses varied according to the policy instrument and the households' characteristics, particularly their lot size. The price elasticity is higher (between –0.29 and –0.47, i.e., more elastic) during the high season (June–October), and lower (between 0 and –0.19, i.e., more inelastic) during the low season (November–May). The voluntary and mandatory conservation programs reduced water use between 1–13% and between 21–29%, respectively, depending on the season and the lot size. The achieved reduction in consumption is however very similar across households. With respect to welfare, households with the smallest lot sizes and lowest income bore the largest proportion of the conservation burden.

Campbell *et al.* (2004) studied the impacts of price and non-price policies implemented by the city of Phoenix, Arizona, during 1990–1996. The authors found that as price apply with high certainty to all households within the service area, even with small price elasticities, the overall water savings may be large. Interestingly, the authors found that inappropriately designed policy can often increase water consumption by off-setting behaviour. Although it was assumed that people who volunteered to pick up water saving retrofit devices would adopt more water saving behaviour compared to those who effortlessly found these devices free on their doorsteps, the results showed increases in water consumption due to rebound effects. Similarly, following the Phase 1 of the Low-Flow Fixtures and Devices Ordinance, water consumption increased by a small amount as people considered this to be their last chance to purchase non-saving devices. However, the results were negative for the consecutive phases and overall the ordinance reduced water consumption by 3.5%. One-to-one communication seemed effective in improving the off-setting effects; however, although the resulting savings per household was large, the overall effect of direct communication was small.

The following paragraphs discuss the general assumptions and findings about the efficacy, cost effectiveness, efficiency and equity of the various price and non-price water demand management policies discussed in this book.

6.1 EFFECT ON WATER DEMAND

Figure 6.1 illustrates the effect of price and non-price approaches on water demand using simple demand-supply curves. The initial market equilibrium is at point $\boldsymbol{a}$, where the current water price ($\boldsymbol{P_0}$) intersects with the existing demand curve $\boldsymbol{D}$ to lead to $\boldsymbol{Q_0}$ as the quantity demanded. When the short run supply of water is restricted to $\boldsymbol{S}$, as a result of droughts for example, the government or water utility may implement price or non-price policies to curb the excess demand. If the price is raised to $\boldsymbol{P_1}$, there will be an upward movement along the demand curve and a new equilibrium will be established at ***b***. The new quantity demanded will be equal to Q_1. If a mandatory restriction is imposed instead, there will be a shift in the demand curve from ***D*** to D_1 and the equilibrium will shift to ***c***, leading to consumption equal to Q_1 again.

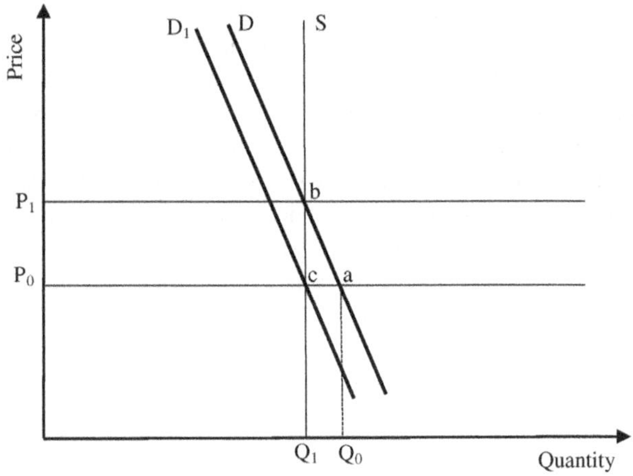

Figrue 6.1 Effect of price and non-price tools on demand (Renwick & Archibald, 1998).

The extent to which water consumption decreases in response to a price increase is determined by the price elasticity of demand for water of households and many authors have attempted to obtain a numerical value for this (see Section 4.4.1). Espey *et al.* (1997) conducted a meta-analysis of 124 estimates generated between 1963 and 1993, and obtained a median value of –0.38 for the short run and –0.64 for the long run, giving an average price elasticity if –0.51. Similarly, in a meta-analysis of almost 300 price elasticity studies between 1963–1998, Dalhuisen *et al.* (2003) obtained a mean value of –0.41. For prices to be effective in reducing water use, the demand should be relatively elastic and water prices should be high enough to make a difference in monthly expenditures. Hence, low income households who spend a greater percentage of their monthly expenditures on water are more sensitive to water price changes than high income households. Similarly, indoor uses which are essential, have a more inelastic demand compared to outdoor uses such as lawn watering. As mentioned above, water demand is more elastic in the long run, as households have the scope to change their behaviours and appliances to reduce demand. In the long run, such price hikes also provide stronger incentives for technological innovation that reduce water consumption (Olmstead & Stavins, 2009).

With price based approaches the certainty of demand reduction is low but level of compliance is high, as the only way to avoid rising prices is to go 'off meter' illegally. On the other hand, non-price based approaches such as mandatory installation of low flow fixtures or ban on outdoor watering seem to have higher certainty of reduced water consumption and lower compliance as monitoring and enforcement would require household inspections (Weitzman, 1974; Olmstead & Stavins, 2009). However, while low flow fixtures may promise to reduce demand, they have often been accompanied by a 'rebound effect' of wasteful consumption practices that lower the expected savings. For example, when the initial models of dual-flush toilets were released, people had to flush the toilet more than once to remove the waste adequately. Similarly, people may take longer showers when their showerheads are replaced with low-flow ones. In a field trial with randomly selected households which had replaced their top-loading washing machines with front-loading ones, Davis (2007) found that clothes washing increased by 5.6%, perhaps because of the cost savings associated with increased efficiency.

6.2 COST-EFFECTIVENESS AND EFFICIENCY

An approach is said to be efficient when its benefits exceed its costs. The design and implementation of water conservation measures, both price and non-price, entail a number of cost elements such as direct costs for fixture labelling and rebates and indirect costs associated with administration, enforcement and monitoring. The benefits, as discussed in Chapter 1, include saving precious water resources, postponing or avoiding facilities expansion, cost savings by individual consumers and environmental benefits. Hence, a comprehensive assessment of efficiency requires a rigorous cost-benefit analysis. Although monetary costs and benefits may be easy to calculate, it is often difficult to put economic values on environmental resources. Thus, policy makers often use cost-effectiveness as a decision making criteria, that is, they seek to achieve the same results at the least cost.

Olmstead and Stavins (2009) explain that although price increases seem to be costly for individual consumers, they are more cost effective for the society as a whole, compared to regulatory policies. Figure 6.2 shows the graphs for the water demand of two different households, which have the same elasticity for indoor uses, but different for outdoor uses as household B is more willing to pay than household A. Initially, the outdoor water consumption for household A is Q_A^{unreg} and household B is Q_B^{unreg} and the indoor water consumption for both is Q_C. A regulatory ban on outdoor water uses forces the demands for household A and B to decrease to Q_A^{reg} and Q_B^{reg} respectively and creates a 'shadow price' (λ) that is higher for household B than for A under the current marginal price ($\boldsymbol{P}$), as B is willing to pay more than A for an additional unit of water. However, a price increase to $\boldsymbol{P}^*$ will achieve the same overall reduction but eliminate the deadweight losses (DWL). Thus, a price approach is more suitable for dealing with heterogeneous demands, as it allows users to adjust their demand according to their preferences.

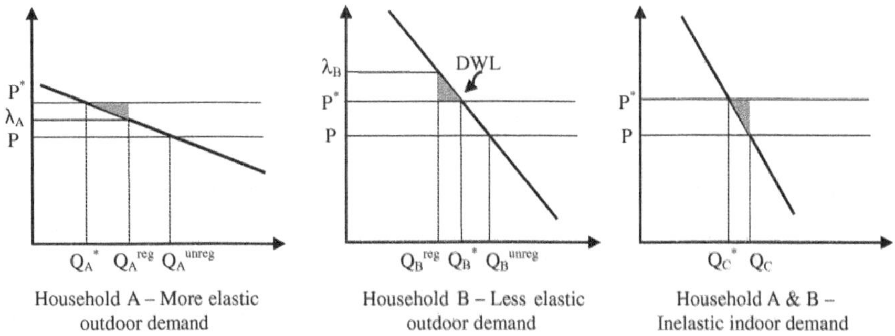

Figure 6.2 Economic losses from outdoor consumption restrictions with heterogeneous outdoor demand (Olmstead & Stavins, 2009).

The fact that the economic losses can be theoretically minimised by using price based approaches instead of non-price based ones have been demonstrated by recent studies. Grafton and Ward (2008) showed that mandatory water restrictions in Sydney, Australia over a single year in 2004–2005 resulted in economic losses of AUD 235 million, or about AUD 150 per household, amounting to half of the average Sydney household water bill in that year.

Moreover, the low price of water reduces the financial gains from installing low flow appliances for which the initial investments may be high. For example, in 1996–1997, in Canberra, the water rate was AUD 0.309/m^3 until a consumption volume of 350 m^3/year, beyond which the rate was AUD 0.721/m^3. Thus, a typical consumer using 350 m^3/year would have a total annual water use bill of around AUD 110, excluding the fixed access charges. Under such circumstances, replacing a conventional 11 litre flush toilet with a dual flush one would cost around AUD 300, save around 40 m^3 per annum and even at the higher block rate of AUD 0.721/m^3, would save around AUD 29 per annum. At a zero discount rate, this would take 10 years to cover costs (Barrett, 2004).

6.3 EQUITY AND DISTRIBUTIONAL EFFECTS

Achieving equity means that low income households do not face a disproportionate burden of the rising costs compared to high-income ones. As price increases are always viewed as an additional burden on the poor, the equity issues can be solved using IBT structure, which enables the poor to meet their basic needs at low cost, but penalises the rich for using more water, especially for outdoor uses. However, the unintended effects of an IBT structure are also well documented (see Section 3.2.2). Poor consumers often have larger household size than richer ones and as each additional person contribute to extra water consumptions, large households are often penalised under this IBT structure. On the other hand, a regulatory ban on outdoor watering poses a disproportionate burden on households which have large gardens.

Renwick and Archibald (1998) studied the distributional effects of both price and non-price demand side management policies in Goleta and Santa Barbara communities in California. During 1989–1990, Santa Barbara first moved from a constant unit price to an IBT and then moved to a much steeper IBT, while Goleta moved from a moderate IBT to a high constant unit price. Goleta's change in pricing policy was intended to support its allocation policy by making billing simpler and easily comprehensible. Goleta implemented a mandatory allocation policy, under which each household was allocated 132 hundred cubic feet (hcf) (374 m^3) per year plus 55% of 1984–88 annual usage above 132 hcf, with penalties of four to ten times the marginal price based on the frequency with which a household exceeded its allotment. On the other hand, Santa Barbara implemented a ban on landscape irrigation except for hand watering and drip systems. The two communities also distributed free low flow showerheads and provided rebates for low-flow toilets. With regard to the distributional effects of price changes, Renwick and Archibald (1998) found that a 10% increase in price would reduce demand by 5.3% and 1.1% for low income and high income households respectively, as price elasticities vary by income class. In terms of regulations, Goleta's allocation policy reduced demand by 7.24 and 3.40 hcf, while Santa Barbara's restriction reduced demand by 6.69 and 4.52 hcf for low density (i.e., larger landscaped areas) and high density households respectively. Similarly, the predicted demand reduction due to adoption of water efficient irrigation was 31% for low density households and 10% for high density ones.

6.4 MONITORING AND ENFORCEMENT

Compared to non-price demand management measures, prices have the advantage of lower administrative costs involved with monitoring and enforcement. As long as consumers are willing to stay connected to the water supplier, they have to comply with the tariff schedule. However, in case of non-price mechanisms, such as mandatory outdoor watering restrictions, it is difficult to ensure compliance. Law enforcing officers will have to regularly patrol all the households in the supply area to look out of non-compliance and fines have to be imposed as penalties.

6.5 POLITICAL AND REVENUE CONSIDERATIONS

Price increases are often not politically popular and are likely to face resistance from households. Olmstead and Stavins (2009) describes this phenomenon as 'fiscal illusion', where households have been so used to the prevalence of subsidised water, that they may object more strongly to water price increases than to increases in other taxes that governments might use to raise revenues. Unlike price increases which raise revenues for the utilities as well, non-price demand management techniques can create political liabilities in the form of water utility budget deficits, because they decrease revenues if the measures are successful.

For example, during a prolonged drought, households in Los Angeles reduced water use in response to their utility's request for voluntary water use reductions. This translated into a fall in total revenues by more than 20%, compelling the utility to implement a rate increase later to cope with its losses (Hall, 2000).

6.6 CONCLUSION

Research has shown that the concurrent implementation of a number of water conservation policies have greater synergistic impact than the use of one policy at a time (Geller *et al.*, 1983). The literature reviewed in this book show that generally price changes and mandatory restrictions or allocation have the greatest impact of water consumption. Technological solutions have the potential of large amounts of water savings, but may result in rebound effects if the policies are not designed appropriately. Public information campaigns are generally effective when used in conjunction with other demand management tools. No matter what policy is implemented, the ultimate aim is to reduce demand by altering attitudes and water use behaviour within the households. The effects of any policy intervention may decline over time. For example, reduction in water demand due to a price increase may not sustain in the long term, unless prices are increased regularly to match inflation and income levels. Similarly, water restrictions can be effective in curbing demand in the short-run, but demands may go back to original levels once restrictions are phased out. These highlight the fact the urban water conservation policies cannot be static and need to be revised from time to time to match the changes in socio-demographical, economic and environmental contexts. As it is believed in Singapore, whether or not there is a water crisis today, people should constantly be reminded of the need to conserve water, so that the future is water secure.

References

Adelman I. and Yeldan A. E. (1999). The Role of Government in Economic Development. Working Paper No. 890, University of California, Berkeley.

Aetra (2013). Fact Sheet [Online]. http://www.aetra.co.id/uploads/fact_sheet_2013/Lembar_Fakta_Jan13.pdf (accessed 16 February 2013)

Agthe D. E., Billings R. B. and Dworkin J. M. (1988). Effects of rate structure knowledge on water use. *Water Resources Bulletin*, **24**(3), 627–630.

Ajzen I. (1991). The theory of planned behavior. *Organisational Behavior and Human Decision Processes*, **50**, 179–211.

Al-Husseini Al Sheikh A. F. (1996). Water and Sanitation in Islam. World Health Organisation, Regional Office for the Eastern Mediterranean, Alexandria, Egypt.

American Society of Mechanical Engineers (2006). ASME A112.19.19–2006 Vitreous China Nonwater Urinals. American Society of Mechanical Engineers, New York.

Arbués F., García-Valiñas M. Á. and Martinez-Espiñeira R. (2003). Estimation of residential water demand: a state-of-the-art review. *Journal of Socio-Economics*, **32**, 81–102.

AsiaisGreen (2008). Photos of Marina Barrage [Online]. http://www.asiaisgreen.com/2008/11/03/photos-of-marina-barrage/ (accessed 19 November 2013)

Atallah S., Khan M. and Malkawi M. (1999). Water conservation through Islamic public awareness in the Eastern Mediterranean Region. *Eastern Mediterranean Health Journal*, **5**, 785.

Bakker K. (2010). Privatizing Water: Governance Failure and the World's Urban Water Crisis. National University of Singapore Press, Singapore.

Bakker K. and Kooy M. (2012). Governance failure: Urban water and conflict in Jakarta, Indonesia. In: Urban Water Conflicts, B. Barraque (ed.), UNESCO and Taylor & Francis, Meppel, The Netherlands, pp. 195–219.

Balasubramanian V. (2013). Bangalore Water Crisis, the Big Hoodwink. DNA (5 March, 2013) [Online]. http://www.dnaindia.com/analysis/1807556/report-bangalore-water-crisis-the-big-hoodwink (accessed 22 June 2013)

Bangalore Water Supply and Sewerage Board (BWSSB) (2013a). Water Tariff and Pro-rata Charges [Online]. http://bwssb.org/water-tarrif-prorata/ (accessed 18 January 2013)

Bangalore Water Supply and Sewerage Board (BWSSB) (2013b). Water Connection [Online]. http://bwssb.org/water-connection/ (accessed 22 June 2013)

Barberán R. and Arbués F. (2009). Equity in domestic water rates design. *Water Resources Management*, **23**, 2101–2118.

Barrett G. (2004). Water conservation: the role of price and regulation in residential water consumption. *Economic Papers: A Journal of Applied Economics and Policy*, **23**, 271–285.

Baumann D. D., Boland J. J. and Sims J. H. (1984). Water Conservation: the struggle over definition. *Water Resource Research*, **20**(4), 428–434, doi: 10.1029/WR020I004P00428

Beal C. and Stewart R. A. (2011). South East Queensland Residential End Use Study, Final Report. Urban Water Security Research Alliance Technical Report No. 47 [Online]. http://www.urbanwateralliance.org.au/publications/UWSRA-tr47.pdf (accessed 05 March 2013)

Beal C., Stewart R. and Huang T.-T. A. (2010). South East Queensland Residential End Use Study: Baseline Results – Winter 2010 Queensland: Urban Water Security Research Alliance.

Beal C., Stewart R., Huang T. and Rey E. (2011). SEQ residential end use study. *Journal of the Australian Water Association*, **38**, 80–84. http://www.manuelectronics.com.au/pdfs/Bealetal_AWA_2011.pdf (accessed 22 June 2013)

Beecher J. A., Mann P. C., Hegazy Y. and Stanford J. D. (1994). Revenue Effects of Water Conservation and Conservation Pricing: Issues and Practices [Online]. National Regulatory Research Institute. http://www.ipu.msu.edu/library/pdfs/publications/Beecher-Mann-Hegazy-NRRI%20Revenue%20Effects.pdf (accessed 28 March 2013)

Bennear L. S., Lee J. M. and Taylor L. O. (2013). Municipal rebate programs for environmental retrofits: An evaluation of additionality and cost-effectiveness. *Journal of Policy Analysis and Management*, **32**, 350–372.

Berk R. A., Cooley T. F., Lacivita C., Parker S., Sredl K. and Brewer M. (1980). Reducing consumption in periods of acute scarcity: the case of water. *Social Science Research*, **9**, 99–120.

Berk R. A., Schulman D., McKeever M. and Freeman H. E. (1993). Measuring the impact of water conservation campaigns in California. *Climatic Change*, **24**, 233–248.

Berkholz P., Stamminger R., Wnuk G., Owens J. and Bernarde S. (2010). Manual dishwashing habits: an empirical analysis of UK consumers. *International Journal of Consumer Studies*, **34**, 235–242.

Berkholz P., Kobersky V. and Stamminger R. (2013). Comparative analysis of global consumer behaviour in the context of different manual dishwashing methods. *International Journal of Consumer Studies*, **37**, 46–58.

Blackburn A. M. (1978). Management strategies: dealing with droughts. *Journal of American Water Works Association*, **70**, 51–59.

Blokker E., Vreeburg J. and Van Dijk J. (2009). Simulating residential water demand with a stochastic end-use model. *Journal of Water Resources Planning and Management*, **136**, 19–26.

Boland J. J. (1997). Pricing Urban Water: Principles and Compromises [Online]. John Hopkins University Available: http://ucowr.org/files/Achieved_Journal_Issues/V92Pricing_Urban_Water.pdf (accessed 20 December 2013).

Boland J. J. (1997). Pricing urban water: principles and compromises. In: Paper Presented at the World Bank Seminar on Pricing of Sanitation and Water Services, February 18–19, 1997.

Bougadis J., Adamowski K. and Diduch R. (2005). Short-term municipal water demand forecasting. *Hydrological Processes*, **19**, 137–148.

Brandez O. M., Renzetti S. and Stinchcombe K. (2010). Worth Every Penny: A Primer on Conservation-Oriented Water Pricing [Online]. http://www.allianceforwaterefficiency.org/uploadedFiles/Resource_Center/Library/rates/POLIS-Primer-on-Conservation-Rate-Structures-May-2010.pdf (accessed 26 March 2013)

Brennan D., Tapsuwan S. and Ingram G. (2007). The welfare costs of urban outdoor water restrictions. *Australian Journal of Agricultural and Resource Economics*, **51**, 243–261.

Brown C. (2012). Water Conservation in the Professional Car Wash Industry. International Carwash Association [Online]. http://www.carwash.org/docs/default-document-library/water-conservation-in-the-professional-car-wash-industry-.pdf?sfvrsn=0 (accessed 15 December 2013)

Bruvold W. H. (1979). Residential response to urban drought in central California. *Water Resources Research*, **15**, 1297–1304.

Budds J. and McGranahan G. (2003). Are the debates on water privatisation missing the point? Experiences from Africa, Asia and Latin America. *Environment and Urbanisation*, **15**(2), 87–114.

Bureau of Metereology (2011). South East Queensland – Water restrictions [Online]. Available: http://www.bom.gov.au/water/nwa/2010/seq/water-restrictions.html (accessed 21 December 2013)

Campbell H. E., Johnson R. M. and Larson E. H. (2004). Prices, devices, people, or rules: the relative effectiveness of policy instruments in water conservation1. *Review of Policy Research*, **21**, 637–662.

Caroma (2013). Profile Smart 305 Round Front Plus [Online]. http://www.caromausa.com/products/index/cu_products/39.php (accessed 20 December 2013)

Carragher B. J., Stewart R. A. and Beal C. D. (2012). Quantifying the influence of residential water appliance efficiency on average day diurnal demand patterns at an end use level: a precursor to optimised water service infrastructure planning. *Resources, Conservation and Recycling*, **62**, 81–90.

CCME Water use Efficiency Task Group (1994). National Action Plan to Encourage Municipal Water use Efficiency [Online]. http://www.ccme.ca/assets/pdf/pn_1168_e.pdf (accessed 22 April 2013)

Centre for International Economics (CIE) (2008). Technical Documents: Updated Estimates of the Cost of Water Restrictions in the ACT Region. Prepared for ActewAGL, September.

Century Research Corporation (1972). *Social Aspects of Urban Water Conservation*. Report prepared for Office of Water Resources Research, U.S. Department of the Interior. NTIS PB-214 970.

Chong J., Kazaglis A. and Giurco D. (2008). Cost Effectiveness Analysis of WELS – the Water Efficiency Labelling and Standards Scheme. Prepared for the Australian Government Department of the Environment, Water, Heritage and the Arts by the Institute for Sustainable Futures [Online], University of Technology, Sydney. Available: http://www.waterrating.gov.au/system/files/publications/2012/10/110/cost-effectiveness-wels.pdf (accessed 20 December 2013)

Clark W. A. and Finley J. C. (2007). Determinants of water conservation intention in Blagoevgrad, Bulgaria. *Society and Natural Resources*, **20**, 613–627.

Clarke J. M. and Brown R. R. (2006). Understanding the factors that influence domestic water consumption within Melbourne. *Australian Journal of Water Resources*, **10**, 261.

Colmar Brunton Social Research (2008). ACT Water Users Research. Final Report, Prepared for Infrastructure Development Branch, Water Division, ActewAGL.

Cooper B., Burton M. and Crase L. (2011). Urban water restrictions: attitudes and avoidance. *Water Resources Research*, **47**, W12527, doi: 10.1029/2010WR010226.

Cooperative Research Centre for Irrigation Futures (CRCIF) (2008). Irrigation of Urban Green Spaces: A Review of the Environmental, Social and Economic Benefits [Online]. IF Technologies Pty Ltd. Available: http://www.irrigationfutures.org.au/imagesDB/news/CRCIF-TR0408-web-final.pdf (accessed 20 December 2013)

Cox J. (2010). IPART's Approach to Urban Water Pricing in NSW [Online]. https://www.google.com.sg/url?sa=t&rct=j&q=&esrc=s&source=web&cd=1&cad=rja&ved=0CDAQFjAA&url=http%3A%2F%2Fwww.aares.org.au%2FCMDownload.aspx%3FContentKey%3Db406d79c-d009–4257-ae24-de3d23f204f2%26ContentItemKey%3D2f9f0529-a3cb-4106-8d7e-ea16a2b176ac&ei=qTpZUerFCpDJrQf5o4CADA&usg=AFQjCNF-o186edHWpPq9kzmmDolKstjUJA&sig2=UixTPo16Pu5_gRW-Pepkcw&bvm=bv.44442042,d.bmk (accessed 26 March 2013)

Cummings S. (2009). Drainline and System Implications of Reduced Flows from High Efficiency Fixtures. WaterSmart Innovations 2009.

Dake K. (1991). Orienting dispositions in the perception of risk an analysis of contemporary worldviews and cultural biases. *Journal of Cross-Cultural Psychology*, **22**, 61–82.

Dake K. (1992). Myths of nature: culture and the social construction of risk. *Journal of Social Issues*, **48**, 21–37.

Dalhuisen J. M., Florax R. J., De Groot H. L. and Nijkamp P. (2003). Price and income elasticities of residential water demand: a meta-analysis. *Land Economics*, **79**, 292–308.

Darby S. (2006). The Effectiveness of Feedback on Energy Consumption. A Review for DEFRA of the Literature on Metering, Billing and direct Displays, 486, 2006.

Davis L. W. (2007). Durable Goods and Residential Demand for Energy and Water: Evidence from a Field Trial [Online]. University of Michigan. Available: http://faculty.haas.berkeley.edu/ldavis/cw.pdf (accessed 20 December 2013)

De Oliver M. (1999). Attitudes and inaction a case study of the manifest demographics of urban water conservation. *Environment and Behavior*, **31**, 372–394.

DEFRA (2010). The New EU Energy Label Explained [Online]. Available: https://www.gov.uk/government/uploads/system/uploads/attachment_data/file/69295/pb13466-eu-energy-label.pdf (accessed 20 December 2013).

Department of Energy (2010). 10 CFR Part 430 Energy Efficiency Program for Consumer Products: Waiver of Federal Preemption of State Regulations Concerning the Water Use or Water Efficiency of Showerheads. Faucets, Water Closets and Urinals; Final Rule. Federal Register.

Department of Mines and Geology (2011). Groundwater Hydrology and Groundwater Quality in and Around Bangalore City [Online]. http://www.indiawaterportal.org/sites/indiawaterportal.org/files/Groundwater_hydrology_and_groundwater_quality_in_and_around_Bangalore_city_DMG_GoK_2011_0.pdf (accessed 21 June 2013)

Department of Singapore Statistics (2012). Singapore in Statistics [Online]. http://www.singstat.gov.sg/pubn/reference/sif2012.pdf (accessed 21 June 2013)

Department of Statistics Singapore (2011). Key Household Characteristics and Household Income Trends [Online]. Department of Statistics Singapore, Singapore. http://www.singstat.gov.sg/pubn/papers/people/pp-s18.pdf (accessed 07 December 2012)

Dhaka Water Supply and Sewerage Authority (DWASA) (2007). Management Information Report for the month of May 2007. Computer Centre, DWASA.

Dolnicar S. and Hurlimann A. (2010). Australians' Water Conservation Behaviours and Attitudes.

Douglas M. and Wildavsky A. B. (1983). Risk and Culture: An Essay on the Selection of Technological and Environmental Dangers. University of California Pr., Berkeley and Los Angeles, California.

Dunlap R. E. and Van Liere K. D. (1978). The new environmental paradigm: a proposed measuring instrument and preliminary results. *The Journal of Environmental education*, **9**, 19.

Edwards K. and Martin L. (1995). A methodology for surveying domestic water consumption. *Journal of the Institution of Water & Environmental Management*, **9**(5), 477–488.

Energy Star (2013). Clothes Washers Key Product Criteria [Online]. Available: http://www.energystar.gov/index.cfm?c=clotheswash.pr_crit_clothes_washers (accessed 19 November 2013)

Espey M., Espey J. and Shaw W. D. (1997). Price elasticity of residential demand for water: a meta-analysis. *Water Resources Research*, **33**, 1369–1374.

Evidence from a Field Trial. University of Michigan [Online]. Available: http://faculty.haas.berkeley.edu/ldavis/cw.pdf (accessed 20 December 2013).

Falkenmark M., Lundquist J. and Widstrand C. (1989). Macro-scale water scarcity requires micro-scale approaches: aspects of vulnerability in semi-arid development. *National Resource Forum*, **13**, 258–267.

Fielding K., Anneliese S., Sally R., Aditi M., Rod M. and John G. (2012). Water Demand Management: Interventions to Reduce Household Water Use. Urban Water Security Research Alliance, Technical Report No. 94 [Online]. Available: http://www.urbanwateralliance.org.au/publications/UWSRA-tr94.pdf (accessed 5 April 2013)

Fielding K., Russell S. and Grace R. (2010). Residential Water Demand Management in South East Queensland: A Report on Water Conservation Beliefs. Urban Water Security Research Alliance Technical Report No. 24.

Financial Times (1999). Making A Big Splash: With more Than a Century in the Private Sector, French Water Companies are Showing the Rest of the World How to Manage Their Water Utilities. Financial Times (24 August 1999).

Fishbein M. and Ajzen I. (1975). Belief, Attitude, Intention, and Behavior: An Introduction to Theory and Research, Addison-Wesley, Reading, MA.

Food and Agricultural Organisation (FAO) (2011). Hot Issues: Water Scarcity [Online]. http://www.fao.org/nr/water/issues/scarcity.html

Garcia-Valinas M. (2006). Analysing rationing policies: drought and its effects on urban users' welfare. *Applied Economics*, **38**, 955–965.

Gaudin S. (2006). Effect of price information on residential water demand. *Applied Economics*, **38**(4), 383–393.

Gauley B. and Koeller J. (2010). Sensor-Operated Plumbing Fixtures: Do They Save Water?: California Energy Commission.

Geller E. S., Erickson J. B. and Buttram B. A. (1983). Attempts to promote residential water conservation with educational, behavioral and engineering strategies. *Population and Environment*, **6**, 96–112.

Gentry B. S. and Auyuan A. T. (2000). Global Trends in Urban Water Supply and Waste Water Financing and Management: Changing Roles for the Public and Private Sectors. Organisation for Economic Co-operation and Development, Paris.

Gilbert J. B. (1978). The California drought—out of disaster, better water management. *Journal of American Water Works Association*, **70**(2), 79–83.

Gilg A. and Barr S. (2006). Behavioural attitudes towards water saving? Evidence from a study of environmental actions. *Ecological Economics*, **57**, 400–414.

Gleick P. (2000a). The changing water paradigm: A look at twenty-first century water resources development. *Water International*, **25**(1), 127–138.

Gleick P. (2000b). The World's Water: The Biennial Report on Freshwater Resources 2000–2001. Island Press, Washington, DC.

Gleick P. (2002). The World's Water: The Biennial Report on Freshwater Resources 2002–2003. Island Press, Washington, DC.

Gleick P. (2003). Soft path's solution to 21st-century water needs. *Science*, **320**, 1524–1528.

Global Partnership on Output-Based Aid (2009). Improved Access to Water Services for Poor Households in Metro Manila [Online]. https://www.gpoba.org/sites/gpoba.org/files/GPOBA%20Manila%207-29-09%20screen.pdf (accessed 25 January 2013)

Global Water Intelligence (GWI) (2012). Tariff Survey [Online]. http://www.globalwaterintel.com/tariff-survey/ (accessed 21 January 2013)

Gordon J., Chapman R. and Blamey R. (2001). Assessing the options for the Canberra water supply: an application of choice modelling. In: The Choice Modelling Approach to Environmental Valuation, J. Bennett and R. Blamey (eds), Edward Elgar, Cheltenham, UK, pp. 73–92.

Grafton R. Q. and Ward M. B. (2008). Prices versus rationing: marshallian surplus and mandatory water restrictions. *Economic Record*, **84**, S57–S65.

Gray O. (1985). Ecological Beliefs and Behavior. Greenwood, Westport, CT.

Griffin R. and Mjelde J. (2000). Valuing water supply reliability. *American Journal of Agricultural Economics*, **82**(2), 414–426.

Grønhøj A. (2006). Communication about consumption: a family process perspective on 'green'consumer practices. *Journal of Consumer Behaviour*, **5**, 491–503.

Gronwall J. (2013). Water access strategies and importance of groundwater for poor urban households in Bangalore, India. Report Submitted to the Institute of Water Policy, National University of Singapore.

Hall D. C. (2000). Public choice and water rate design. *The Political Economy of Water Pricing Reforms*, 189–212. https://getinfo.de/app/Public-Choice-and-Water-Rate-Design/id/BLCP%3ACN035849658

Hardin, G. (1968). The Tragedy of the Commons. Science, 162, 1243-1248.

Hanemann W. M. (1993). Designing New Water Rates for Los Angeles. University of California, Berkeley [Online]. http://opensiuc.lib.siu.edu/cgi/viewcontent.cgi?article=1421&context=jcwre (accessed 28 March 2013)

Hanemann W. M. and Nauges C. (2005). Heterogeneous Responses to Water Conservation Programs: The Case of Residential Users in Los Angeles [Online]. Department of Agriculture and Resource Economics and Policy, University of California at Berkeley.

Available: http://ageconsearch.umn.edu/bitstream/7158/2/wp061026.pdf (accessed 30 December 2013)

Hanson A.H. (1959). Public Enterprise and Economic Development. Routledge & Kegan Paul, London.

Hardin G. (1968). The Tragedy of the commons. *Science*, **162**, 1243–1248.

Harland P., Staats H. and Wilke H. A. (1999). Explaining proenvironmental intention and behavior by personal norms and the theory of planned behaviorl. *Journal of Applied Social Psychology*, **29**, 2505–2528.

Hensher D., Shore N. and Train K. (2006). Water supply security and willingness to pay to avoid drought restrictions. *Economic Record*, **82**, 56–66.

Heste K. J.-V. (2012). Top-Loading vs Front-Loading Washers – Which is Better? [Online]. Available: http://housewares.about.com/lw/Home-Garden/Home-improvement-renovation/Top-Loading-vs-Front-Loading-Washers-Which-is-Better-.htm (accessed 19 November 2013)

Hills S., Birks R. and Mckenzie B. (2002). The Millennium Dome "Watercycle" experiment: to evaluate water efficiency and customer perception at a recycling scheme for 6 million visitors. *Water Science & Technology*, **46**, 233–240.

Hoehn, J. P. (2011). Economic Principles of Water Conservation Tariffs and Incentives [Online]. University of Michigan. Available: http://cdn.intechopen.com/pdfs/24683/InTech-Economic_principles_for_water_conservation_tariffs_and_incentives.pdf (accessed 28 March 2013)

Hoque S. F., Asanga G. and Mafizur R. (2014). Chapter 7: Potential for residential water conservation in Dhaka. In: Sustainability Matters: Volume 2: Asia's Energy Concerns, Green Policies and Environmental Advocacy, L-H. Lye, R. S. Victor, M. C. Loke, E. Y. Liya and K. Harn-Wei (eds), World Scientific Publishing, Singapore (Forthcoming).

Hoque S. F. and Wichelns D. (2013). State-of-the-art review: designing urban water tariffs to recover costs and promote wise use. *International Journal of Water Resources Development*, **29**(3), 472–491.

How to Save Water (2013). Water Efficient Showers [Online]. Available: http://www.how-to-save-water.co.uk/ (accessed 19 November 2013)

Howe C. and Smith M. (1994). The value of water supply reliability in urban water systems. *Journal of Environmental Economics and Management*, **26**, 19–30.

Inglehart R. (1995). Public support for environmental protection: objective problems and subjective values in 43 societies. *PS: Political Science and Politics*, **28**, 57–72.

Inglehart R. (1977). The Silent Revolution: Changing Values and Political Styles among Western Publics. Princeton University Press, Princeton, NJ.

Inman D. and Jeffrey P. (2006). A review of residential water conservation tool performance and influences on implementation effectiveness. *Urban Water Journal*, **3**, 127–143.

Institute for Sustainable Futures and ACIL Tasman (2009). Review of Water Restrictions. Prepared for the National Water Commission.

Institute of Water Policy (2013). Comparative Study of Water and Wastewater Tariff Structures. National University of Singapore, Singapore.

International Conference on Water and the Environment (ICWE) (1992). The Dublin Statement on Water and Sustainable Development [Online]. http://www.wmo.int/pages/prog/hwrp/documents/english/icwedece.html (accessed 30 August 2013)

Jacobs H. (2008). Residential water information management. *South African Journal of Informafion Management*, **10**, 12.

Jakarta Water Supply Regulatory Body (2009). The First Ten Years of Implementation of the Jakarta Water Supply 25-Year Concession Agreement (1998–2008) [Online]. http://www.jakartawater.org/images/stories/unduh/10tahunbrEng.pdf (accessed 25 January 2013)

Johannesburg Water (2010). Water and Sanitation Tariffs 2010/11 – JW Annexure A 2011 Update [Online]. http://www.johannesburgwater.co.za/ (accessed 28 March 2013)

Kantola S., Syme G. and Campbell N. (1982). The role of individual differences and external variables in a test of the sufficiency of Fishbein's model to explain behavioral intentions to conserve water. *Journal of Applied Social Psychology*, **12**, 70–83.

Kathmandu Upatyaka Khanepani Limited (2009). Annual Operating Report. 1 Falgun 2064-31 Asadh 2065 [Online]. http://www.kathmanduwater.org/reports/Annual%20 Operating%20Report(FY64_65).pdf (accessed 12 February 2013)

Koss P., and Khawaja M. (2001). The value of water supply reliability in California: a contingent valuation study. *Water Policy*, **3**, 165–174.

Kurz T. (2002). The psychology of environmentally sustainable behavior: fitting together pieces of the puzzle. *Analyses of Social Issues and Public Policy*, **2**, 257–278.

LADWP (2010). Urban Water Management Plan Los Angeles Department of Water and Power.

Lahlou A. A. and Attia B. (2005). Managing Water Demand: Policies, Practices and Lessons from the Middle East and North Africa Forums. IWA Publishing and International Development Research Centre (IDRC), Cornwall, UK.

Lam S. P. (1999). Predicting intentions to conserve water from the theory of planned behavior, perceived moral obligation, and perceived water right. *Journal of Applied Social Psychology*, **29**, 1058–1071.

Lam S. P. (2006). Predicting intention to save water: theory of planned behavior, response efficacy, vulnerability, and perceived efficiency of alternative solutions. *Journal of Applied Social Psychology*, **36**, 2803–2824.

Little A. (1996). Bangkok Water Supply Demand Side Management Study. Metropolitan Waterworks Authority, Bangkok, Thailand.

Los Angeles Department of Water and Power (2012a). Water Rates [Online]. https://www.ladwp.com/ladwp/faces/ladwp/aboutus/a-financesandreports/a-fr-waterrares?_adf.ctrl-state=vspn0fuvq_79&_afrLoop=348018122901000 (accessed 15 November 2012)

Los Angeles Department of Water and Power (2012b). Water Rates Ordinance No. 170435 [Online]. https://www.ladwp.com/cs/idcplg?IdcService=GET_FILE&dDoc Name=A D17DWPWEB9173007496&RevisionSelectionMethod=LatestReleased (accessed 16 February 2013)

Lovins A. (1976). Energy Strategy: The Road Not Taken? 55 Foreign Affairs 65 (1976–1977) [Online]. http://heinonline.org/HOL/LandingPage?collection=journals&handle=hein.journals/fora55&div=8&id=&page= (accessed 30 August 2013)

Luan I. O. B. (2010). Singapore water management policies and practices. *International Journal of Water Resources Development*, **26**(1), 65–80.

Luz J. M. and Paladio-Melosantos M. L. (2012). Manila, Philippines. In: Good Practices in Urban Water Management: Decoding Good Practices for A Successful Future, S. K. Anand Chiplunkar (ed.), Asian Development Bank and National University of Singapore, Manila and Singapore.

Mainieri T., Barnett E. G., Valdero T. R., Unipan J. B. and Oskamp S. (1997). Green buying: The influence of environmental concern on consumer behavior. *The Journal of Social Psychology*, **137**, 189–204.

Maki J. E., Hoffman D. M. and Berk R. A. (1978). A time series analysis of the impact of a water conservation campaign. *Evaluation Quarterly*, **2**, 107–18.

Manila Water (2011). Annual Report [Online]. http://www.manilawater.com/investor/Investor%20Resources/Pages/DownloadableMaterials.aspx (accessed 16 February 2013)

Manila Water (2012). Water Tariff with Effect from January 2013 [Online]. http://www.manilawater.com/investor/Corporate%20Governance/Pages/SECFillings.aspx (accessed 16 February 2013)

Marcotullio, P. and McGranahan, G. (eds) (2007). Scaling Urban Environmental Challenges: From Local to Global and Back. Earthscan, UK and USA.

Martin N. (1999). Population, Households and Domestic Water Use In Countries of the Mediterranean Middle East (Jordan, Lebanon, Syria, the West Bank, Gaza and Israel). International Institute for Applied Systems Analysis [Online]. Available: http://webarchive.iiasa.ac.at/Admin/PUB/Documents/IR-99-032.pdf (accessed 20 December 2013)

Martin W. E. and Thomas J. F. (1986). Policy relevance in studies of urban residential water demand. *Water Resources Research*, **22**, 1735–1741.

Martínez-Espiñeira R. and García-Valiñas M. Á. (2012). Adopting versus adapting: adoption of water-saving technology versus water conservation habits in Spain. *International Journal of Water Resources Development*, **29**(3), 1–15.

Masten S. E. (2004). Public utility ownership in 19th-century America: the 'aberrant' case of water. *The Journal of Law, Economics and Organisation*, **27**(3), 604–654.

Mayer P. W., Deoreo W. B., Opitz E. M., Kiefer J. C., Davis W. Y., Dziegielewski B. and Nelson J. O. (1999). Residential End Uses of Water. American Reasearch Foundation and American Water Works Association Research Foundation (AWWARF), Denver, Colorado, USA [Online]. http://www.waterrf.org/PublicReportLibrary/RFR90781_1999_241A.pdf (accessed 30 August 2013)

Mercer L. J. and Morgan W. D. (1980). Impact of a water conservation campaign: Some extensions on a time series analysis. *Evaluation Review*, **4**, 107–117.

Michelson A. M., Mcguckin J. T. and Stumpf D. (1999). Nonprice water conservation programs as a demand management tool. *Journal of the American Water Resources Association*, **35**, 593–602.

Midden C. J., Kaiser F. G. and Teddy Mccalley L. (2007). Technology's four roles in understanding individuals' conservation of natural resources. *Journal of Social Issues*, **63**, 155–174.

Millock K. and Nauges C. (2010). Household adoption of water-efficient equipment: the role of socio-economic factors, environmental attitudes and policy. *Environmental and Resource Economics*, **46**, 539–565.

Ministry of Information, Communications and the Arts (MICA) (2003). Water Talks? if Only it Could. Singapore.

National University of Singapore (NUS) (2013). Temasek Foundation Water Leadership Programme (TFWLP) [Online]. http://lkyspp.nus.edu.sg/executive-education/open-enrolment-programmes/tfwlp/ (accessed 15 December, 2013)

National Water Commission (2011). Efficient water resource pricing in Australia: an assessment of administered scarcity pricing in urban areas, Waterlines Report Series No 44 [Online]. http://www.nwc.gov.au/data/assets/pdf_file/0016/10933/44_Scarcity.pdf (accessed 28 March 2013)

Nauges C. and Whittington D. (2010). Estimation of water demand in developing countries – an overview. *World Bank Research Observation*, **25**(2), 263–294. doi: 10.1093/wbro/lkp016

Navya P. K. (2013). Why Cauvery Stage IV Water Has Reached Bangalore, But Not Your Home. Citizen matters (7 March 2013) [Online]. http://bangalore.citizenmatters.in/articles/print/4988-why-stage-iv-cauvery-water-has-reached-bangalore-but-not-your-home (accessed 22 June 2013)

Nieswiadomy M. L. (1992). Estimating urban residential water demand: effects of price structure, conservation, and education. *Water Resources Research*, **28**(3), 609–615.

North Carolina Department of Environment and Natural Resources (2009). Recommendations for Water Efficiency Standards for Water-Using Fixtures in Residential and Commercial Buildings.

NYC Environmental Protection (2013). About Automated Meter Reading (AMR) [Online]. http://www.nyc.gov/html/dep/html/customer_services/amr_about.shtml (accessed 28 March 2013)

OECD (2009). Managing Water for All: An OECD Perspective on Pricing and Financing [Online]. http://www.oecd.org/greengrowth/sustainable-agriculture/44476961.pdf (accessed 28 March 2013)

Office of Water Regulation (OFWAT) (1993). Privatisation and History of the Water Industry. Information Note No. 18.

OFWAT (2011a). Household Data: Metered and Unmetered Household Numbers for Water 2010–11 [Online]. http://www.ofwat.gov.uk/regulating/reporting/rpt_tar_2010–11householddata (accessed 28 January 2013)

OFWAT (2011b). Push, Pull, Nudge – How Can We Help Customers Save Water, Energy and Money? [Online]. http://www.ofwat.gov.uk/publications/focusreports/prs_inf_pushpullnudge.pdf (accessed 28 November 2013)

Olmstead S. M. and Stavins R. N. (2009). Comparing price and nonprice approaches to urban water conservation. *Water Resources Research*, **45**, W04301.

Osann E. R. and Young J. E. (1998). Saving Water, Saving Dollars: Efficient Plumbing Products and the Protection of America's Waters.

Palyja (2011). Annual Report [Online]. http://en.palyja.co.id/__wysiwyg/filemanager/files/Publication/PALYJA%20Annual%20Report%202011%20-%20web.pdf (accessed 25 January 2013)

Palyja (2012). Water for All [Online]. Available: http://en.palyja.co.id/main-business/water-for-all/ (accessed 25 January 2013)

PAM Jaya (2007). Water Tariff Rate [Online]. http://en.palyja.co.id/uploads/Tarif.pdf (accessed 25 November 2012)

Postel S. (1998). Water for food production: will there be enough in 2025? *BioScience*, **48**(8), 629–637.

Postel S. (1999). Pillar of Sand: Can the Irrigation Miracle last? Norton/Worldwatch, New York, USA.

Postel S. (2001). Safeguarding our water – Growing more food with less water. *Scientific American*, **284**(2), 40–45.

Potter J. (1994). Dilemmas in Water and Wastewater Pricing: Case Study of Bangkok, Masters thesis, submitted to the Department of Urban Studies and Planning, MIT, Thailand.

Prasad N. (2007). Privatisation of water: a historical perspective. *Law, Environment and Development Journal*, **3**(2), 217.

Productivity Commission (2011). Australia's Urban Water Sector, Report No. 55, Final Inquiry Report, Canberra.

Programs: The Case of Residential Users in Los Angeles. Department of Agriculture and Resource Economics and Policy [Online], University of California at Berkeley. Available: http://ageconsearch.umn.edu/bitstream/7158/2/wp061026.pdf (accessed 30 December 2013).

Public Utilities Board (PUB) (1976). Annual Report. PUB, Singapore.

Public Utilities Board (PUB) (1978). Annual Report. PUB, Singapore.

Public Utilities Board (PUB) (1979). Annual Report. PUB, Singapore.

Public Utilities Board (PUB) (1981). Annual Report. PUB, Singapore.

Public Utilities Board (PUB) (1983). Annual Report. PUB, Singapore.

Public Utilities Board (PUB) (1986). Annual Report. PUB, Singapore

Public Utilities Board (PUB) (1987). Annual Report. PUB, Singapore.

Public Utilities Board (PUB) (1995). Annual Report. PUB, Singapore.

Public Utilities Board (PUB) (2011a). Water Efficiency Labelling Scheme [Online]. Available: http://www.pub.gov.sg/wels/Pages/default.aspx (accessed 10 January 2013)

Public Utilities Board (PUB) (2011b). Water Conservation Awareness Programme 2011 [Online]. http://www.pub.gov.sg/conserve/WACProgramme/Pages/default.aspx (accessed 15 December, 2013)

Public Utilities Board (PUB) (2011c). Water Efficiency Labelling Scheme [Online]. Singapore. Available: http://www.pub.gov.sg/wels/rating/Pages/Requirements.aspx (accessed 15 April 2013)

Public Utilities Board (PUB) (2013a). Water Conservation Awareness Programme 2013 [Online]. Available: http://www.pub.gov.sg/conserve/WACProgramme/Pages/2013Initiatives.aspx (accessed 15 December 2013)

Public Utilities Board (PUB) (2013b). The Singapore Water Story: From Vulnerability to Strength [Online]. http://www.pub.gov.sg/water/Pages/singaporewaterstory.aspx (accessed 30 August 2013)

Public Utilities Board (PUB) (2013c). Domestic helpers [Online]. Singapore. Available: http://www.pub.gov.sg/conserve/Households/Pages/Domestichelpers.aspx (accessed 15 August 2013)

Queensland Water Commission (2010). South East Queensland – Water Strategy [Online]. Available: http://www.dews.qld.gov.au/data/assets/pdf_file/0019/32734/seqws.pdf (accessed 21 December 2013)

Randolph B. and Troy P. (2008). Attitudes to conservation and water consumption. *Environmental Science and Policy*, **11**, 441–455.

Renwick M. and Archibald S. (1998). Demand side management policies for residential water use: who bears the conservation burden? *Land Economics*, **73**, 343–359.

Renwick M. and Green R. (2000). Do residential demand side management policies measure up? An analysis of eight Californian water agencies. *Journal of Environmental Economics and Management*, **40**, 37–55.

Richter C. P. (2010). In-house Consumer Study on Dishwashing Habits in Four European Countries: Saving Potentials in Households with Dishwashing Machine. Schriftenreihe der Haushaltstechnik Bonn, 2, Shaker Verlag, Aachen.

Rogers P., Bhatia R. and Huber A. (1998). Water as A Social and Economic Good: How to Put the Principle into Practice [Online]. Global Water Partnership/Swedish International Development Cooperation Agency, Stockholm, Sweden. http://info.worldbank.org/etools/docs/library/80637/IWRM4_TEC02-WaterAsSocialEconGood-Rogers.pdf (accessed 24 January 2013)

Roth G. (1987). The Private Provision of Public Services in Developing Countries. Oxford University Press, Oxford.

Rubin S. J. (2003). The Cost of Water and Wastewater Service in the United States. National Rural Association, Duncan.

Russell S. and Fielding K. (2010). Water demand management research: a psychological perspective. *Water Resources Research*, **46**, W05302.

Sammer K. and Wüstenhagen R. (2006). The influence of eco-labelling on consumer behaviour–Results of a discrete choice analysis for washing machines. *Business Strategy and the Environment*, **15**, 185–199.

Sarafidis V. and Sydney Water (2011). The Residential Price Elasticity of Demand for Water [Online]. http://www.pc.gov.au/data/assets/pdf_file/0004/106537/sub083.pdf (accessed 12 March 2013)

Schleich J. and Hillenbrand T. (2009). Determinants of residential water demand in Germany. *Ecological Economics*, **68**, 1756–1769.

Schuck E. and Profit R. (2004). Evaluating non-price water demand policies during severe droughts. In: Selected Paper Prepared for the American Agricultural Economics Association Annual Meeting, July 1–4, Denver, Colorado.

Schlunke A., Lewis J., and Fane S. (2008). Analysis of Australian Opportunities for More Water-Efficient Toilets. Institute of Sustainable Futures [Online]. Available: http://www.waterrating.gov.au/system/files/publications/2012/10/108/water-efficient-toilets.pdf (accessed 20 December 2013).

Schwartz S. H. (1973). Normative explanations of helping behavior: a critique, proposal, and empirical test. *Journal of Experimental Social Psychology*, **9**, 349–364.

Schwartz S. H. (1977). Normative influences on altruism1. *Advances in Experimental Social Psychology*, **10**, 221–279.

Sciences. Pennsylvania State University [Online]. Available: http://pubs.cas.psu.edu/FreePubs/pdfs/uh164.pdf (accessed 20 December 2013)

Scott D. and Willits F. K. (1994). Environmental attitudes and behavior a Pennsylvania survey. *Environment and Behavior*, **26**, 239–260.

Scottish Water (2012). Metered Household Charges 2012–2013 [Online]. http://www.scottishwater.co.uk/you-and-your-home/your-charges/2012–2013-charges/information-about-your-charges/metered-household-charges-201213-leaflet (accessed 24 January 2013)

Seattle Public Utilities (2011). Annual Survey of Wholesale Customers: Summary of Results. Seattle Public Utilities. Available: http://www.operatingboard.org/groups/public/@spu/documents/webcontent/01_015563.pdf (accessed 30 December 2013)

Sharpe W. and Swistock B. (2008). Household Water Conservation [Online]. College of Agricultural Sciences. Pennsylvania State University. Available: http://pubs.cas.psu.edu/FreePubs/pdfs/uh164.pdf (accessed 20 December 2013)

Shiklomanov I. (1999). State Hydrological Institute (Shl. St. Petersburg) and United Nations Educational, Scientific and Cultural Organisation. UNESCO, Paris.

Singapore Power (SP Services) (2013a). Average Water Consumption by Household Type [Online]. http://www.singaporepower.com.sg/irj/servlet/prt/portal/prtroot/docs/guid/609ffede-040d-2f10-f5b4-de464caacd39?spstab¼Our%20Services (accessed 12 March 2013)

Singapore Power (SP Services) (2013b). Electricity Tariff Revision for the Period 1 July to 30 September 2013 [Online]. http://www.singaporepower.com.sg/irj/go/km/docs/wpccontent/Sites/SP%20Services/Site%20Content/Tariffs/documents/latest_press_release.pdf (accessed 07 July 2013)

Spiller D. (2008). Water for today, water for tomorrow: establishment and operation of the SEQ water grid. *Australian Economic Review*, **41**, 420–427.

Standards Australia (2005). ATS 5200.004–2005 Technical Specification for Plumbing and Drainage Products Part 004: Urinal Flushing Cisterns. Standards Australia, Sydney.

Standards Australia/Standards New Zealand (2005). AS/NZS 6400: 2005 Water Efficient Products – Rating and Labelling. Standards Australia/Standards New Zealand, Sydney and Wellington.

Stern P. C. (2000). New environmental theories: toward a coherent theory of environmentally significant behavior. *Journal of Social Issues*, **56**, 407–424.

Stern P. C., Dietz T., Abel T., Guagnano G. A. and Kalof L. (1999). A value-belief-norm theory of support for social movements: the case of environmentalism. *Human Ecology Review*, **6**, 81–98.

Stevens T. H., Miller J. and Willis C. (1992). Effect of price structure on residential water demand. *Water Resources Bulletin*, **28**, 681–685.

Stewart R. A., Willis R. M., Panuwatwanich K. and Sahin O. (2013). Showering behavioural response to alarming visual display monitors: longitudinal mixed method study. *Behaviour & Information Technology*, **32**, 695–711.

Syme G. J., Nancarrow B. E. and Seligman C. (2000). The evaluation of information campaigns to promote voluntary household water conservation. *Evaluation Review*, **24**, 539–578.

Taipei Water Department (2011). Water Charges [Online]. http://english.twd.gov.tw/ct.asp?xItem=994144&CtNode=23944&mp=114012 (accessed 16 October 2012)

Takizawa S., Iwasaki C. and Oguma K. (2005). Effects of water tariff structures on water demand in Tokyo metropolis. *Water Science and Technology: Water Supply*, **5**(6), 235–242.

Tan Y. S., Lee T. J. and Tan K. (2009). *Clean, Green and Blue: Singapore's Journey Towards Environmental and Water Sustainability*. Ministry of Environment and Water Resources, ISEAS Publishing, Singapore.

Thames Water (2012). Metered Charges 2012–13 [Online]. http://www.thameswater.co.uk/tw/common/downloads/literature-water-waste-water-charges/201213-metered-charges-leaflet.pdf (accessed 16 October 2012)

The Hindu (2013). Borewell Owners Have Time Till March 31 to Register. The Hindu (10 March 2013) [Online]. http://www.thehindu.com/news/national/karnataka/borewell-owners-have-time-till-march-31-to-register/article4492061.ece (accessed 22 June 2013)

The Telegraph (2011). Power showers use twice as much water and electricity as baths [Online]. 22 November 2011. Available: http://www.telegraph.co.uk/earth/

energy/8904423/Power-showers-use-twice-as-much-water-and-electricity-as-baths.html (accessed 19 November 2013)

Thøgersen J. and Ölander F. (2002). Human values and the emergence of a sustainable consumption pattern: A panel study. *Journal of Economic Psychology*, **23**, 605–630.

Tortajada C. and Joshi Y. K. (2013). Water demand management in Singapore: involving the public. *Water Resources Management*, **27**(8), 2729–2746.

Tortajada C., Joshi Y. and Biswas A. K. (2013). The Singapore Water Story: Sustainable Development in an Urban City-state. Routledge, London and New York.

Trumbo C. W. and O'Keefe G. J. (2001). Intention to conserve water: environmental values, planned behavior, and information effects. A comparison of three communities sharing a watershed. *Society &Natural Resources*, **14**, 889–899.

U.S. EPA (2007). High-Efficiency Lavatory Faucet Specification Version 1.0 [Online]. Available: http://www.epa.gov/WaterSense/docs/faucet_spec508.pdf (accessed 19 November 2013)

U.S. EPA (2008). National Efficiency Standards and Specifications for Residential and Commercial Water-Using Fixtures and Appliances [Online]. http://www.epa.gov/watersense/docs/matrix508.pdf (accessed 19 November 2013)

U.S. EPA (2010). WaterSense Specification for Showerheads Version 1.0 [Online]. Available: http://www.epa.gov/WaterSense/docs/showerheads_finalspec508.pdf (accessed 19 November 2013)

U.S. EPA (2013). WaterSense – Rebate Finder [Online]. Available: http://www.epa.gov/owm/water-efficiency/rebate_finder_saving_money_water.html (accessed 27 November 2013)

UNESCAP (2002). Public Awareness on Water Conservation and Water Efficiency [Online]. http://www.unescap.org/esd/Energy-Security-and-Water-Resources/water/publications/water/wrs/81.asp (accessed 24 July 2013)

UNILEVER (2011). UK Sustainable Shower Study [Online]. Available: http://www.unilever.co.uk/media-centre/pressreleases/2011/sustainableshowerstudy.aspx (accessed 19 November 2013)

United Nations (UN) (1992). Chapter 18 – Protection of the Quality & Supply of Freshwater Resources: Application of Integrated Approaches to the Development, Management & Use of Water Resources. Agenda 21 [Online]. http://www.un.org/esa/dsd/agenda21/res_agenda21_18.shtml (accessed 30 August 2013)

United Nations (UN) (2000a). Resolution Adopted by the General Assembly 54/175: The Right to Development [Online]. http://www.un.org/en/ga/search/view_doc.asp?symbol=A/RES/54/175&Lang=E (accessed 30 August 2013).

United Nations (UN) (2000b). Millennium Development Goals [Online]. http://www.un.org/millenniumgoals/environ.shtml (accessed 30 August 2013)

United Nations (UN) (2007). Report of the United Nations High Commissioner for Human Rights on the Scope and Content of the Relevant Human Rights Obligations Related to Equitable Access to Safe Drinking Water and Sanitation Under International Human Rights Instruments [Online]. http://daccess-dds-ny.un.org/doc/UNDOC/GEN/G07/136/55/PDF/G0713655.pdf?OpenElement (accessed 30 August 2013)

United Nations (UN) (2010). Resolution Adopted by the General Assembly 64/292: The Human Right to Water and Sanitation [Online]. http://www.un.org/en/ga/search/view_doc.asp?symbol=A/RES/64/292 (accessed 30 August 2013)

United Nations (UN) (2011a). The Human Right to Water and Sanitation: Milestones [Online]. http://www.un.org/waterforlifedecade/pdf/human_right_to water_and_sanitation_milestones.pdf (accessed 30 August 2013)

United Nations (UN) (2011b). The Human Right to Water and Sanitation: Media Brief [Online]. http://www.un.org/waterforlifedecade/pdf/human_right_to_water_and_sanitation_media_brief.pdf (accessed 30 August 2013)

United Nations, Office of the High Commissioner for Human Rights (OHCHR) (2002). Committee on Economic, Social and Cultural Rights – General Comments. General Comment No. 15 [Online]. http://www.unhchr.ch/tbs/doc.nsf/0/a5458d1d1bbd713fc1256cc400389e94/$FILE/G0340229.pdf (accessed 30 August 2013)

United Utilities (2012). Our charges 2012–2013 [Online]. http://www.unitedutilities.com/Our-charges-2012-2013.aspx (accessed 16 October 2012)

Vickers A. (2001). Handbook of Water Use and Conservation. WaterPlow Press, Amherst Massachusetts.

Villaraigosa A. R. (2008). Securing L.A.'s Water Supply. City of Los Angeles Department of Water and Power.

Vogler S. (2013). Southeast Queensland Water Restrictions Lifted but Higher Use Will Cost Residents. The Courier-mail, 02 January.

Wang Y. -D., Song J.-S., Byrne J. and Yun S.-J. (1999). Evaluating the persistence of residential water conservation: a 1992–1997 panel study of a water utility program in Delaware. *Journal American Water Resources Association*, **35**(8), 1269–1276.

Water Supplies Department (2011). Annual Report 2010–2011 [Online]. http://www.wsd.gov.hk/filemanager/common/annual_report/2010_11/index_en.htm (accessed 30 August 2013)

Weitzman M. L. (1974). Prices vs. *quantities. The Review of Economic Studies*, **41**, 477–491.

Which? (2013). Water saving products [Online]. Available: http://www.which.co.uk/energy/creating-an-energy-saving-home/reviews-ns/water-saving-products/ (accessed 19 November 2013)

Whittington, D. (1992). Possible adverse effects of increasing block water tariffs in developing countries. Economic Development and Cultural Change, 41, 75–87.

Whittington D. (2003). Municipal water pricing and tariff design: a reform agenda for South Asia. *Water Policy*, **5**, 61–76.

Wichelns, D. (2013). Enhancing the performance of water prices and tariff structures in achieving socially desirable outcomes. *International Journal of Water Resources Development*, **29**(3), 310–326.

Williams A., Whitehead C. D. and Lutz J. (2011). A Systems Framework for Assessing Plumbing Products-Related Water Conservation. Lawrence Berkeley National Laboratory [Online]. Available: http://eetd.lbl.gov/sites/all/files/a_systems_framework_lbnl-5286e.pdf (accessed 10 December 2013)

Willis R., Stewart R. A., Panuwatwanich K., Capati B. and Giurco D. (2009). Gold Coast domestic water end use study. *Water: Journal of the Australian Water Association*, **36**, 79–95.

Willis R. M., Stewart R. A., Panuwatwanich K., Williams P. R. and Hollingsworth A. L. (2011). Quantifying the influence of environmental and water conservation attitudes on household end use water consumption. *Journal of Environmental Management*, **92**, 1996–2009.

Winpenny J. (2003). Financing Water for All. 3rd World Water Forum, World Water Council and Global Water Partnership [Online]. http://www.worldwatercouncil.org/fileadmin/world_water_council/documents_old/Library/Publications_and_reports/CamdessusReport.pdf (accessed 30 August 2013)

World Bank (2012). World Development Indicators [Online]. http://databank.worldbank.org/ddp/home.do?Step=1&id=4 (accessed 30 August 2013)

World Water Council (2000). Ministerial Declaration of the Second World Water Forum [Online]. http://www.idhc.org/esp/documents/Agua/Second_World_Water_Forum%5B1%5D.pdf (accessed 30 August 2013)

World Water Council (2003). Ministerial Declaration of the Third World Water Forum [Online]. http://www.mofa.go.jp/policy/environment/wwf/declaration.html (accessed 30 August 2013)

Worthington A. C. and Hoffman M. (2008). An empirical survey of residential water demand modelling. *Journal of Economic Surveys*, **22**, 842–871.

Yamanishi O. and Manabe H. (2002). Residential water consumption by activity in general household. In: Proceeding of 46th Research meeting of Japan Water Works Association's Kansai District, 61–64 (in Japanese).

IWA Publishing's authorised EU representative for General Product Safety Regulations is Diane D'Arras, 15 rue Duret, 75116 Paris, France, e-mail: safety@iwap.co.uk.

Printed and bound by CPI Group (UK) Ltd, Croydon, CR0 4YY
12/05/2026
02108880-0006